Statehouse
AND
Greenhouse

Statehouse
AND
Greenhouse

The Emerging
Politics of American
Climate Change Policy

Barry G. Rabe

BROOKINGS INSTITUTION PRESS
Washington, D.C.

Copyright © 2004
THE BROOKINGS INSTITUTION
1775 Massachusetts Avenue, N.W., Washington, D.C. 20036
www.brookings.edu

Library of Congress Cataloging-in-Publication data
Rabe, Barry George, 1957–
  Statehouse and greenhouse : the stealth politics of American climate change policy / Barry G. Rabe.
    p.   cm.
  Includes bibliographical references and index.
    ISBN 0-8157-7310-2 (cloth : alk. paper) —
    ISBN 0-8157-7309-9 (pbk. : alk. paper)
    1. Climatic changes—Government policy—United States. 2. Climatic changes—Government policy—Economic aspects—United States. I. Title.
  QC981.8.C5R33 2004
  363.738'7456'0973—dc22   2003026257

9 8 7 6 5 4 3 2 1
The paper used in this publication meets minimum requirements of the American National Standard for Information Sciences—Permanence of Paper for Printed Library Materials: ANSI Z39.48-1992.

Typeset in Sabon

Composition by OSP, Inc., Arlington, Virginia

Printed by R.R. Donnelley, Harrisonburg, Virginia

To Dana

Contents

Preface

One of the finest graduate students I have ever taught approached me several years ago in considerable distress. She had left a promising job, with a corporation celebrated nationally for its commitment to environmental protection, to begin work on a master's degree in environmental policy. After three semesters of study, she was deeply frustrated. "It all seems so pointless," she explained. "In class after class, reading after reading, we hear gloom and doom about every area of the environment. If it really is so bad and so hopeless, why bother?"

Perhaps no environmental issue triggers such feelings of hopelessness as global climate change. Measurement of greenhouse gases, including carbon dioxide, show a steady level of increase over the latter part of the twentieth century. In the century and a half during which temperatures have been monitored worldwide, the fifteen warmest years on record have occurred since 1970. In North America, the pace of ice-cap melting and sea-ice collapse has accelerated in recent years, corresponding with similar developments in Antarctica. A number of nations—as well as regions of the United States—have experienced a range of unusually damaging weather episodes in recent years. Much recent analysis of the possible effects of climate change resembles a litany of horrors of biblical proportions: massive flooding, disease outbreaks, loss of habitat, destruction of species, and escalating illness and death owing to extreme temperatures.

Intense scholarly debate continues over the connections between these events and rising levels of greenhouse gases. Analysis becomes even more contentious as scientists from diverse disciplines—and ideological perspectives—attempt to project future trends. Slight adjustments to exceedingly complex models of climate change can render dramatically different results. Any imprecision from the natural and physical sciences is compounded by social science and policy analysis, resulting in a dizzying array of economic and social interpretations of the possible impacts of global warming.

Consequently, much of the best-known analysis on this subject entails extreme forecasts and hyperbolic prose. Many of the leading books on global warming tend to offer long-term scenarios that project ecological doomsday in short order, barring truly heroic international commitment to reducing greenhouse gases. In *Dead Heat,* the environmentalist Michael Oppenheimer and the journalist Robert Boyle present an account of an America that makes the *Grapes of Wrath* seem mild by comparison. Rapid climate change leaves deserted ghost towns from Colorado to Indiana, blizzards of topsoil darken the skies across North America, waves of migration have moved millions from the United States to Canada and Mexico, and massive food shortages and poverty have left the nation on the verge of collapse. This book was published in 1990 as a forecast of how American life would change by 2000, followed by predictions of an even grimmer set of developments in subsequent decades, including a desperate move of the American capital from Washington, D.C., to the most northern reaches of Michigan's Upper Peninsula.[1] Their narrative has been matched by comparable accounts in recent years and even joined by heavily publicized movies such as Kevin Costner's *Water World* and a television miniseries, CBS's *The Fire Next Time.* All project a world spinning out of control because indolent politicians failed to respond to early warnings of imminent catastrophe.

These extreme accounts have triggered the creation of a cottage industry of writers and analysts who lurch to the opposite extreme, dismissing global climate change as, at most, a minor inconvenience requiring modest adaptation. In fact, such analyses foresee economic and social catastrophe only if proposed international policy remedies, such as the Kyoto Protocol, are implemented. The economist Thomas Gale Moore has argued that implementation of such policies "would lead to worldwide recession, rising unemployment, civil disturbances, and increased tension between nations as accusations of cheating and violations of international treaties inflamed passions."[2] Moore and others contend that projected global warming would actually improve the quality of life in the United

States, proclaiming that significant temperature increases by the end of the current century would boost the American economy by more than $100 billion a year. In turn, a 2003 report by the Chicago-based Heartland Institute projects that state government efforts to reduce greenhouse gases will "impose unbearable burdens on state treasuries and on consumers and businesses."[3] Calling for the repeal of virtually all existing programs, the institute wildly inflates previous worst-case economic scenarios of Kyoto implementation. In so doing, it predicts that serious state reduction efforts will trigger a state fiscal catastrophe of a magnitude comparable to the level of ecological degradation anticipated by Oppenheimer and Boyle as a result of failure to implement such policies.

Such interpretations, far-fetched as they may be, tend to dominate media accounts and national policy discourse, only compounding the extreme difficulties in forging reasonable policy responses to the challenge of global climate change. Unlike more conventional environmental problems, climate change cannot be addressed through regulations concentrated on a small set of industries or nations. Indeed, many of the tools used to confront more conventional problems, such as air and water pollution, may be poorly suited to climate change, given the ubiquity of sources that produce carbon dioxide, methane, and other greenhouse gases. Grander regulatory schemes, perhaps best reflected in the Kyoto Protocol, are conceptually engaging. They are, however, riddled with loopholes and face daunting implementation problems, even if a larger percentage of the world's nations were willing to participate. "The burden to be shared is large, there are no accepted standards of fairness, nations differ greatly in their dependence on fossil fuels, and any regime to be taken seriously has to promise to survive a long time," notes the economist Thomas Schelling. "The precedents are few."[4]

It is perhaps no surprise, therefore, that an American federal government known for its penchant for institutional gridlock has accomplished virtually nothing in this area in the past decade. This policy inertia could itself serve as fodder for another blockbuster movie. But that is not the story of this book. Alongside the extremist rhetoric and the barriers to serious engagement of this issue at the federal level, an almost stealth-like process of policy development has been evolving. State governments and their policies that influence greenhouse gases rate barely a mention in most leading scholarly or press accounts of the evolution of American climate-change policy. Nevertheless, over the past decade approximately one-third of the American states have enacted multiple policies that show considerable promise of reducing greenhouse gases. Another sizable cluster of states have launched similar experimentation, if on a smaller scale. These new policies are neither merely symbolic steps nor the grander stuff of international emissions-trading schemes. Nonetheless, they demonstrate that it is politically possible in the United States both to form coalitions to support initiatives to reduce greenhouse gases and to take initial steps to secure implementation.

Collectively, these policies may point to an alternative policy architecture for addressing climate change, at least for the approximately 25 percent of worldwide greenhouse gases that emanate each year from the United States. They demonstrate innovation in virtually every policy sector relevant to the generation of greenhouse gas, including energy, air quality, transportation, agriculture, and natural resources. In some instances, these policies stem from long-standing state experience in—and jurisdiction over—a particular policy area. Electricity restructuring has been an obvious area for innovation, drawing on decades of state regulatory experience and the pressure to reinvent the industry in response to deregulation and the emergence of new technologies for generating electricity. In other cases, they reflect markedly new departures for any American government, national or subnational. These range from negotiated agreements for greenhouse gas reductions by specific industries to programs that disclose carbon dioxide emission levels to the state and general citizenry.

This book is intended to examine the evolution of state government policies on global climate change. It has four primary goals. First, it introduces a diverse set of innovative cases, offering a detailed description of some of the kinds of policies to combat global warming that are now being put into place in the United States. In subsequent chapters, twelve states receive intensive review, reflecting considerable economic, political, and regional diversity and thereby providing an overview of current state practice. These cases demonstrate a range of responses, reflecting tremendous state-by-state variance. On the one hand, New Jersey, New Hampshire, Oregon, and Wisconsin have addressed the climate change issue with vigor, developing a series of reduction policies and demonstrating a level of policy sophistication that may rival the staunchest European nations supporting the Kyoto Protocol. On the other hand, states such as Louisiana have proved largely indifferent—despite the vast potential threat from anticipated sea-level rise—whereas still others, such as Michigan, have taken steps to deter state officials from taking any action to reduce greenhouse gases. Another cluster of states falls somewhere between these camps, although the clear national trend is toward greater state action. Rather than a study that concentrates exclusively on success stories, this overview is intended to consider the full range of state government policy responses demonstrated to date.

Second, the book attempts to explain why state innovation in global climate change has been relatively vigorous and why so little attention has been drawn to it thus far. Most analysis of American climate-change policy is written as if the United States were a unitary form of government, either ignoring the realities of federalism or dismissing potential subnational roles with a presumption that states have no unilateral incentive to take action on a problem with such global consequences. This study demonstrates that the process of climate-change policymaking is alive and active in many state capitals. This engagement often reflects the capacity of policy entrepreneurs to cultivate new policy ideas and to build coalitions that attain broad support, although much of this work is done quietly and gradually. Such entrepreneurs tend to be based in state government agencies, including those devoted to environmental protection or energy, and often develop a reputation as trusted experts on climate-change science and policy. They frequently draw upon previous experience in related policy areas to develop their policy proposals and frame them as economic development opportunities. They then find coalition partners, including other agency members, elected officials, and members of various industries and environmental groups, to build a broad supportive base. Each case study includes a detailed account of the underlying politics that have either facilitated or precluded policy development in various states, offering collectively a depiction of policy formation that is less dramatic but more functional than conventional accounts of the federal level.

Third, the book attempts to draw larger lessons from this recent flurry of American experience. For the past decade, the federal government has been unable to agree on much of anything about climate change, other than to periodically support more funding for research, most of which is concentrated in the natural and physical sciences. Regardless of the partisan composition of the federal government, little has happened, whether Democrats controlled both executive and legislative branches (as they did from 1993 to 1995), a Democratic president was matched with a Republican Congress (as from 1995 to 2001), or a Republican president served with either a divided or Republican Congress (as from 2001 to 2004). Across these periods, many states with wide variation in partisan control actually moved further than their federal counterparts, both in formulating new policy and in achieving some demonstrable level of greenhouse gas reduction. Is this a fluke, or are there larger lessons—and possibilities—to be gained? Concluding chapters draw on recent state experience to consider possible next steps. These options include state-by-state diffusion of innovation, federal adoption of best state practices, and multistate collaboration that may extend beyond American boundaries. In addition, I consider the viability and desirability of introducing greenhouse gas reduction into a larger process of envisioning a next generation of environmental protection that places greater emphasis on states.

Fourth, the book is intended to contribute to moving debate over global climate change from bombast to the realm of what is politically and technically feasible. As the policy analyst David Victor has noted in his penetrating review of the serious problems in the Kyoto Protocol, "Focusing on action would help to break the odd stasis on global warming policy: the political process has been an oasis for goal setters yet a desert of action."[5] My analysis contends that though there has been an increasingly significant amount of "action," it has been concentrated at a governmental level that goes largely ignored by many scholars and commentators. Meaningful strategies to reduce greenhouse gases, in the United States and internationally, will take considerable time, resources, and experimentation. They will need to engage citizens in the process of policy development and implementation rather than imposing structures established during all-night binges of international bargaining.[6] The process may well include a range of policy instruments that are, at present, poorly understood and rarely employed. Ironically, given this kind of challenge, American states may be emerging as international leaders at the very time the national government continues to be portrayed as an international laggard on global climate change.

This analysis is based on a confluence of data sources, many of which were gathered through a series of site visits and interviews. Twelve states were selected for intensive analysis, chosen through a process that attempted to maximize diversity in terms of state policy engagement (or lack thereof) on climate change, physical size and economic composition, levels of greenhouse gas emissions, geographical location, and partisan control. As a result, the study is intended to provide a representative sample of current American state experience and is clearly not designed to select and highlight "best practices" across the various areas relevant to greenhouse gas reduction. Selection was completed after extensive consultation with staff from the Pew Center on Global Climate Change, the National Association of State Energy Officials, the Environmental Council of the States, and the U.S. Environmental Protection Agency. More than one hundred individuals were interviewed for this study, primarily in late 2001 and the first half of 2002. Twelve of these interviews represented follow-ups to interviews conducted in late 1998 and 1999, during a pilot phase of the study.

Interview participants represented diverse sectors, including state and federal government, industry, and environmental advocacy groups. Interviews averaged from thirty minutes to three hours and followed a standard, semistructured interview protocol. Interview participants were assured that there would be no direct attribution of their comments without their permission. Initial plans to tape-record interviews were abandoned in response to concerns expressed about confidentiality by several early participants. I am profoundly grateful to all individuals who consented to be interviewed for this study. The interviews provided invaluable supplements to other primary sources, such as government reports and documents, local analyses of proposed policies, and state legislation and executive orders. There remains remarkably little secondary literature from which to draw, reflecting the limited analysis of state climate-change policy to date, although local media accounts proved helpful in a number of instances.

Funding for this research was provided by the Pew Center on Global Climate Change, the Faculty Studies Research Grant Program of the Canadian Embassy, and the Office of the Provost and Executive Vice President for Academic Affairs of the University of Michigan. It has become increasingly customary in studies of climate-change policy to allege that researchers holding views that differ from the author have, in effect, been "bought" by external funding sources, whether they are private or public. Consequently, I have concluded that any publications on climate change underwritten by external funding should contain some financial disclosure information. In this study, let me be clear about the support that I have received. The majority of this funding was used to hire research assistants, support costs of travel for field visits, and recover indirect costs. During 2001 and 2002, this funding support provided approximately 5 percent of my total personal income. None of these funding sources reviewed the manuscript before publication.

There is, in many respects, a Rip Van Winkle quality to this study. I decided in the mid-1990s that in the event that I was promoted to full professor it would be appropriate to tackle an entirely new area of research and write a book about it. I was marginally literate on the topic of climate change and was intrigued by the interdisciplinary, interjurisdictional, and intersectoral issues that it posed. When the promotion occurred and sabbatical loomed shortly thereafter, climate change seemed the perfect topic. This project began as a pilot study of early American state and Canadian provincial responses to the challenge of greenhouse gas reduction, following the American and Canadian decisions to sign (if not ratify) the Kyoto Protocol. There appeared to be only the beginnings of state strategies evident at this point. In particular, I distinctly remember making one of my first formal presentations on this topic, in early 1999, during which I was interrupted by a distinguished environmental economist who said that there was nothing to this endeavor. In his view, no state government would unilaterally take steps to reduce greenhouse gases without clear marching orders and incentives from federal authorities and international regimes. "There is no rational reason for them to do this," he said. I was not convinced by this dismissal and was instead more eager than ever to examine the potential state and federal interplay on this issue.

This was about the time that I began an eighteen-month scholarly slumber, attributable to my agreement to serve as interim dean of the School of Natural Resources and Environment at the University of Michigan. During this period, I had little chance to sustain the research, though the field I chose to study was indeed changing rapidly. As I began life as a recovering dean with a sabbatical in January 2002, I quickly grasped that the very landscape of American climate-change policy was changing dramatically, with a particularly notable tilt toward expanded and intensified state action and a clear pattern of federal disengagement.

This led me to expand the early goals of the study and necessitated development of a strong network of colleagues to help guide me through this evolving policy landscape. I was particularly fortunate to receive financial support from the Pew Center on Global Climate Change, but I also benefited greatly from close collaboration with Eileen Claussen, Judith Greenwald, Taryn Fransen, and Paige Messec. This collaboration contributed to the publication in November 2002 of a Pew Center report that I wrote. I also received considerable assistance from Andrea Denny, Steven Dunn, Susan Gander, Denise Mulholland, and Julie Rosenberg of the U.S. Environmental Protection Agency, Kirsten Oldenburg of the U.S.

Department of Transportation, and David Terry and Jeff Genzer of the National Association of State Energy Officials.

This network of colleagues helped guide the development of the research design and implementation of the research plan that led to the completion of this book. In addition to the many individuals who consented to extended interviews, I also benefited greatly from colleagues who read either some or all of the manuscript in draft form or provided particularly insightful comments during various public presentations. These include Tom Arrandale, William Becker, Rebecca Blank, John Chamberlin, Terry Davies, John Dernbach, Raymond DeYoung, Alicia Farmer, David Feldman, Martha Feldman, Robert Gasaway, Gail Grella, Michael Greve, Michael Kraft, William Lowry, Philip Mundo, Pietro Nivola, Walter Rosenbaum, David Terry, Edward Weber, and Gregory Yantz. Three external reviewers commissioned by Brookings were extremely helpful, as were three fine research assistants, Katherine Irvine, Alex Belinky, and Matthew Weinbaum. I am also grateful to have had a number of opportunities for public presentation of earlier versions of this work, including venues provided by the American Enterprise Institute, the American Political Science Association, the Association for Canadian Studies in the United States, the International Political Science Association Research Committee on the Structure and Organization of Government, the Kauper Lectureship of Lutheran Campus Ministry of the University of Michigan, the National Conference of State Legislatures, the National Press Club, the University of Michigan Society of Fellows, and the U.S. Environmental Protection Agency. In addition, I am profoundly grateful to colleagues at the Matthaei Botanical Gardens of the University of Michigan for providing me a comfortable and collegial setting in which to write and revise this manuscript.

This marks the publication of my third book with the Brookings Institution Press, and it is a continuing privilege to work with its research and publications staff. Pietro Nivola of the Governance Studies Program has been an especially strong supporter of this project from the outset and has provided tremendous insight about the interplay between energy and environmental protection policy. Christopher Kelaher and Robert Faherty of the Brookings Press have been characteristically professional and a pleasure to work with throughout. Katherine Kimball improved the manuscript greatly through careful editing.

Finally, I would like to acknowledge the three people who have been affected most by the Rip Van Winkle odyssey that culminates in this book.

My sons, Matthew and Andrew, are now of an age to study issues concerning climate change in their own public school settings and are quick with critical comment on everything from televised commercials concerning greenhouse gases to ways in which our family life-style seems inconsistent with the principles set forth in some of the state cases that follow. Their counsel, on all matters, remains invaluable. My wife, Dana Runestad, has read every word of this book and heard the underpinnings of every argument. No one has worked harder to help me think through these issues, encourage me to finish this book in the face of competing duties, or appreciate the much larger context that puts all of this work into perspective. Naturally, given this level of family engagement, it would not be difficult to shift blame for any mistakes to them. However, the normal conventions apply, and any errors are solely my responsibility.

CHAPTER 1

The Politics of Climate Change, State Style

George W. Bush and Christine Todd Whitman may well have signed into law policies that will achieve greater levels of greenhouse gas reduction than those approved by any other elected officials in the United States during the past decade. These policies are in no way associated, however, with the federal offices they assumed in 2001. Instead, their environmental reputations reflect major policies that were approved during their governorships in Texas and New Jersey, respectively, and are currently being implemented by their successors there.

For Bush, this entailed signing the Texas Public Utility Regulatory Act of 1999. This bill included an ambitious program that required Texas utilities to increase their reliance over the next decade on renewable energy sources that do not generate greenhouse gases. It outlined a detailed plan to increase steadily the level of renewable energy used in the state and established penalties for noncompliance. As a result, Texas has experienced a "wind rush" and is expected to generate between 3 and 4 percent of its electricity from renewable wind power by 2010, up from a rate below 1 percent when the legislation was signed. Given the enormous scope of the Texas economy and its heavy reliance on electricity, it is estimated that this legislation will reduce Texas carbon dioxide emissions by 1.83 million metric tons a year by 2009. These reductions may be greater than the entire carbon dioxide (CO_2) emissions generated by either Vermont or the District of Columbia in that year (see table 1-1).

Table 1-1. *Greenhouse Gas Emissions of the United States, by State, 1999*
Million metric tons of carbon equivalent

Rank/state	Emissions	Emissions per million people	Rank/state	Emissions	Emissions per million people
1 Texas	166.56	7.64	27 South Carolina	20.93	5.09
2 California	94.83	2.70	28 Iowa	20.65	2.07
3 Ohio	69.75	6.12	29 Kansas	19.43	7.19
4 Pennsylvania	64.05	5.21	30 Massachusetts	17.16	2.68
5 Florida	60.83	3.62	31 Arkansas	17.09	6.30
6 Indiana	59.85	9.73	32 Mississippi	17.05	5.94
7 Illinois	58.58	4.67	33 Wyoming	16.79	33.91
8 Michigan	52.69	5.25	34 Utah	16.60	7.20
9 New York	52.31	2.75	35 New Mexico	15.10	8.21
10 Louisiana	51.16	11.46	36 North Dakota	13.82	22.02
11 Georgia	43.11	5.02	37 Oregon	11.24	3.19
12 North Carolina	37.19	4.47	38 Nebraska	11.11	6.48
13 Kentucky	36.43	8.91	39 Alaska	11.03	17.16
14 Alabama	35.90	8.01	40 Nevada	10.91	4.94
15 Missouri	35.17	6.21	41 Connecticut	10.09	2.93
16 Tennessee	32.36	5.59	42 Montana	8.37	9.23
17 New Jersey	32.10	3.75	43 Maine	4.86	3.74
18 West Virginia	30.65	17.06	44 New Hampshire	4.55	3.55
19 Virginia	29.62	4.06	45 Delaware	4.30	5.32
20 Wisconsin	27.97	5.14	46 Hawaii	4.25	3.43
21 Oklahoma	25.04	7.20	47 Idaho	4.11	3.05
22 Minnesota	25.02	4.98	48 South Dakota	3.63	4.79
23 Washington	23.11	3.80	49 Rhode Island	3.08	2.88
24 Arizona	21.47	3.91	50 Vermont	1.77	2.87
25 Colorado	21.32	4.70	51 District of Columbia	1.13	1.98
26 Maryland	21.16	3.80			

Source: National Environmental Trust, *First in Emissions, Behind in Solutions* (Washington: National Environmental Trust, 2002); U.S. Environmental Protection Agency, *Inventory of U.S. Greenhouse Gas Emissions and Sinks: 1990–2000* (2002); "Energy CO_2 Inventories" (yosemite.epa.gov/OAR/global-warming.nsf/content/EmissionsInternationalInventory.html [July 24, 2002]) and "International Emissions" (yosemite.epa.gov/OAR/globalwarming.nsf/content/Emissions.html [July 24, 2002]) from the U.S. Environmental Protection Agency's website link to Global Warming.

Bush may have referred to this legislation on the presidential campaign trail—most likely during a speech in Saginaw, Michigan, in September 2000 that featured a widely noted commitment to take steps to reduce greenhouse gases, if elected. What went less noted was a tweak at his main opponent, incumbent vice president Al Gore, for backing away during the campaign from his earlier, more aggressive proposals and instead emphasizing a more modest mixture of tax credits and voluntary reduction programs to reduce greenhouse gases. Bush contended that Texas had found a better method to reduce greenhouse gas emissions. He promised to build on that experience if elected president.

Whitman's involvement included active support through a series of administrative orders and endorsement of legislation involving an across-the-board commitment by the state of New Jersey to reduce its greenhouse gas emissions to 3.5 percent below 1990 levels by 2005. This went well beyond the reduction levels pledged by President George H. W. Bush in the 1992 United Nations Framework Convention on Climate Change that has essentially gone ignored by the United States and most other national signatories. In fact, New Jersey's actions signaled a willingness to reach beyond the 1992 agreements and put the state in line to attain the levels of reduction pledged by the United States in the 1997 Kyoto Protocol—7 percent below 1990 levels by 2012. The state took this action even though Kyoto stood no chance of Senate ratification during the Clinton-Gore administration and was never submitted for Senate consideration.

New Jersey is on track to reach these reduction targets, which have continued to be supported by Whitman's Democratic successor, James McGreevey. The New Jersey strategy has emphasized a series of programs that involve virtually every sector with some impact on greenhouse gas releases, mixing a series of coercive and voluntary regulatory tools. "New Jersey has set an ambitious goal to not only curb greenhouse gas emissions, but to reduce them," explained Whitman in endorsing the reduction pledge. "The fact is that climate change associated with greenhouse gases has an effect on every aspect of our daily lives. The environmental and economic benefits that stem from controlling greenhouse gases are enormous."[1]

The move from the statehouse to the White House led to far more cautious approaches to greenhouse gases by former governors Bush and Whitman. In many respects, the federal government's role in climate change remains as unclear today as it was in the late 1980s, when a convergence of research findings and steamy summers thrust climate change onto the national political agenda. Ironically, this inconclusive soap opera on the Washington stage has served to obscure an increasingly dynamic and active process of developing policies to reduce carbon dioxide, methane, and other greenhouse gases. Dozens of state laws have been enacted—many since 1998—that establish specific state-based strategies with the explicit intent of reducing greenhouse gases. These involve formal commitments in virtually every sector that generates such gases. When combined, they constitute an almost stealth-like approach to global climate change, in that they have received remarkably little attention from scholars, journalists, or environmentalists.

These state programs, of course, lack the political sex appeal of an international trading regime for greenhouse gases. Collectively, they only begin to make a dent in the levels of reduction that may be warranted in coming decades. Moreover, they do not involve all states, some states being either indifferent or hostile toward such policies. Nonetheless, while climate change policy appears hopelessly deadlocked in Washington, a set of state governments that cuts across partisan and regional lines is demonstrating that it is possible to make some significant inroads on this issue, often through creative initiatives tailored to particular state circumstances and opportunities.

Collectively, these policies indicate alternative ways to address global climate change that may be particularly relevant for a nation as physically large and economically diverse as the United States. Indeed, many American states release a higher amount of greenhouse gases each year than many nations that are either Kyoto signatories or prominent candidates for future involvement. Texas, for example, exceeds the United Kingdom and France in annual emissions, just as Ohio exceeds Taiwan, Illinois exceeds Thailand, Georgia exceeds Argentina, New Jersey exceeds Egypt, Wisconsin exceeds Pakistan (see table 1-2), Colorado exceeds Iraq, and Massachusetts exceeds Norway. In fact, if each of the fifty states were to secede from the union and secure national sovereignty, many would rank among the world's leading national sources of greenhouse gases. In turn, states can address greenhouse gas emissions directly or through the many key policy areas they dominate—from electricity regulation to land use—that are profoundly relevant to any long-term effort to reduce greenhouse gases. They have turned increasingly to these powers in recent years, often expanding them in creative ways, in the process of redefining American climate change policy.

The evolution of climate change policy over the past decade from Washington to the states has a number of roots, not the least of which is the enormous policy gap created by federal action—and inaction. In their efforts to bridge this gap, some states have characterized the possibility of early action on climate change as an environmental necessity that could offer economic advantages, others as an economic development opportunity that warrants exploration, and still others as an economic threat to be avoided at all costs. This, in turn, affects the ways in which states choose to label their emerging policies. Obviously, virtually every policy

Table 1-2. *Greenhouse Gas Emissions of the American States and Other National Emissions Leaders, 1999*[a]

Million metric tons of carbon equivalent

Rank/nation or state	Emissions	Rank/nation or state	Emissions
1 United States	1,526.1	29 *Indiana*	59.9
2 China	792.1	30 *Illinois*	58.6
3 Russian Federation	444.1	31 Netherlands	53.1
4 Japan	301.3	32 *Michigan*	52.7
5 India	240.0	33 *New York*	52.3
6 Germany	230.7	34 *Louisiana*	51.1
7 *Texas*	166.6	35 Thailand	45.0
8 United Kingdom	153.6	36 *Georgia*	43.1
9 Canada	153.3	37 Romania	37.4
10 France	126.2	38 *North Carolina*	37.2
11 Italy	122.8	39 *Kentucky*	36.4
12 South Africa	104.8	40 Argentina	36.4
13 South Korea	104.8	41 Venezuela	36.2
14 Mexico	104.1	42 *Alabama*	35.9
15 Ukraine	102.9	43 *Missouri*	35.2
16 Australia	96.0	44 Belgium	33.2
17 *California*	94.8	45 Czech Republic	33.2
18 Brazil	91.8	46 *Tennessee*	32.3
19 Poland	91.3	47 *New Jersey*	32.1
20 Spain	83.8	48 United Arab Emirates	32.0
21 Iran	79.7	49 Egypt	31.7
22 *Ohio*	69.8	50 *West Virginia*	30.7
23 Taiwan	67.4	51 Singapore	30.4
24 Turkey	66.8	52 *Virginia*	29.6
25 Indonesia	66.5	53 Kazakhstan	28.9
26 North Korea	66.6	54 Greece	28.2
27 *Pennsylvania*	64.1	55 *Wisconsin*	28.0
28 *Florida*	60.8	56 Pakistan	27.9

Source: See table 1-1.

a. Individual states of the United States are in italics.

enacted by any government, from closures of military bases to construction of schools, could have some impact on greenhouse gases. Rather than examine such "accidental" policies, this book focuses on two other types of policies. One set includes those policies that expressly attempt to reduce carbon dioxide or other greenhouse gases, such as laws in New Hampshire and Massachusetts that regulate carbon dioxide emissions from electrical utilities as part of a "multipollutant" strategy. Another set includes those policies that are not explicitly labeled as efforts to reduce greenhouse gases but whose proponents are clearly aware of the likely impact and committed to monitoring any reductions. These types of policies include the Texas "renewables portfolio standard" signed into law by Governor Bush and Georgia's crosscutting effort to reduce motor vehicle use in the Atlanta metropolitan area.

The capacity of policy entrepreneurs to advance ideas for greenhouse gas reduction, working within different political and economic contexts, influences what they can and cannot attempt to do. Such entrepreneurs in these cases are most commonly found in the upper tiers of state agencies with jurisdiction over energy, environmental protection, transportation, agriculture, forestry, and natural resources. They work within existing political and resource constraints but nurture ideas and coalitions that lead to broadly supported greenhouse gas reduction policies tailored to the particular features of their state. In particular, successful entrepreneurs prove highly effective at linking greenhouse gas reduction initiatives with long-term economic development opportunities for a state. In many instances, they propose new policies on the basis of previous experience in related areas, such as emissions trading or energy efficiency, and have established themselves as highly credible experts in designing new policies.

The range of recent state experience is explored in an examination of a dozen states in later chapters of this book. These states were selected to maximize partisan, economic, social, and regional diversity. They demonstrate a wide range of possible policy responses, from strong commitment to intense opposition. As a result, they offer a representative blend of recent American state experience rather than an exclusive sampler of best practices and success stories.

The Rocky Road from Rio: Climate Change Science and Politics

At first glance, global climate change would appear to be the quintessential policy problem requiring a top-down strategy imposed by an international regime and implemented through binding agreements with national governments. Greenhouse gases are generated by all of the world's nations, although developed nations such as the United States have been dominant sources. Any long-term strategy to stabilize or reduce greenhouse gas levels will require widespread engagement, including the involvement of developing nations such as China and India that are increasingly large contributors to the problem (see table 1-2). Unlike other environmental problems that can literally be exported from one jurisdiction to another, such as industrial wastes and conventional air pollutants, any release—or reduction—of greenhouse gases from any source has global ramifications.

Global climate change thus constitutes a unique type of environmental problem and challenge to collective action. Whereas most environmental policy constitutes a response to pollutants or substances widely perceived as bad for the environment and for human health, greenhouse gases are truly a mixed bag. Some of these gases, such as carbon dioxide, trap heat in the atmosphere and have fostered temperature levels in recent centuries conducive to staggering rates of agricultural and economic productivity as well as population growth. Many of these gases are familiar to the most basic science student, largely "natural" in origin, as opposed to such conventional "contaminants" as hazardous wastes and toxic air pollutants. Hence they are hard to vilify, especially in the absence of exact cause-and-effect evidence of their impact. Moreover, greenhouse gas releases have been a fairly reliable proxy for overall rates of economic development; the very concept of a "decarbonized" society may be equally disconcerting to developed nations that generate substantial emissions and developing nations eager to expand their economies through increased reliance on fossil fuel.

At the same time, however, greenhouse gas levels have grown at a steady—and alarming—rate in the industrial era. Over the past two centuries, atmospheric levels of carbon dioxide have increased by at least 30 percent, methane levels have more than doubled, and nitrous oxide levels have climbed by 15 to 20 percent.[2] This rate of increase has been particularly dramatic in recent decades, owing to population growth and continuing changes in economic development and transportation. In the event that no policies are implemented to curb current rates of carbon dioxide release, its concentration by 2100 is projected to be at least 50 percent higher than 1995 levels.

The increase in greenhouse gases has corresponded in recent decades with a series of alarming developments concerning climate. Overall, global surface temperature increased by approximately 1 degree Fahrenheit (about 0.6 degrees Celsius) during the twentieth century. The year 2002 was the second-hottest year worldwide since the advent of modern record-keeping in 1880, and nine of the ten highest annual temperatures during this era have been registered since 1990. The National Research Council, drawing heavily on a series of studies published by the Intergovernmental Panel on Climate Change, estimates that projected growth in greenhouse gas levels will produce a global surface temperature rise between

2.5 and 10.4 degrees Fahrenheit (1.4 and 5.8 degrees Celsius) by the end of the current century.[3] In turn, global sea levels rose from ten to twenty-five centimeters over the past hundred years and are projected to further increase by as much as fifty centimeters during the current century.[4]

In North America, warming Pacific Ocean temperatures are contributing to significant declines in fish, mammals, birds, and seaweed from California to Alaska. A number of states with large coastal areas or large surfaces at or below sea level have experienced a range of unusually intensive weather episodes in recent years, particularly southeastern states such as Louisiana and Florida. The likely impact of future changes varies enormously by region, and estimates remain inexact, but no portion of the United States is expected to remain unaffected. Even a chilly state such as Minnesota, often the brunt of economists' jokes anticipating an upswing for the state in tourist visits and retirement relocations as a result of rising temperatures, may face serious challenges.[5] State and national studies project that by 2050 the state will see a tripling of heat-related deaths and that by 2100 it may be contending with a significant dieback of boreal forests, major shifts in soil receptivity to agricultural bounty, possible temperature increases of between 2 and 7 degrees Fahrenheit, the introduction of new diseases, and substantial loss of fish and species habitats.[6]

All of these linkages and projections, of course, remain enormously inexact. Slight adjustments in exceedingly complex climate models can render dramatically different results. Nonetheless, more than a decade and a half ago, the issue of global climate change moved onto the agenda of the federal government and, in less publicized ways, the agendas of many state governments as well. In fact, various periods of federal action—and inertia—on climate change since the late 1980s have had the direct effect of intensifying state action.

The first serious federal steps to address greenhouse gases involved a series of actions approved during a period of intense partisan conflict between a Democratic Congress and a Republican president, George H. W. Bush. These initiatives did not reflect a comprehensive strategy and tended to eschew language such as "greenhouse gases" and "carbon dioxide." But they were at least partly influenced by growing concern over global climate change and provided the first federal infrastructure for subsequent action. They are also particularly noteworthy in that virtually no subsequent legislation of consequence emerged from either the eight years of the Clinton administration or the first three years of the George W. Bush administration, leaving a decade-long gap in federal action that states have clearly begun to fill.

First, the 1992 Energy Policy Act set broad rules for restructuring the delivery of electricity and attempting to reduce American dependence on foreign oil. Perhaps its most important impact has been to give states new latitude in redesigning long-entrenched electricity markets and considering alternatives to fossil fuels. The legislation promoted alternatives through the creation of tax incentives, most notably a tax credit of $0.017 a kilowatt-hour to generators of new sources of renewable electricity from wind, solar, geothermal, and related sources. It also required utilities to report their carbon dioxide emissions to the federal government and established the first voluntary registry whereby greenhouse gas reductions could be reported and given potential credit in any future regulatory regime. These features served as precursors for expanded actions taken by states in more recent years, many of which are examined in subsequent chapters.

Second, the Clean Air Act Amendments of 1990 introduced many states to the practice of emissions trading, through a novel and successful approach to reducing sulfur dioxide emissions that was influential in the formation of the Kyoto Protocol.[7] That experience has prompted states to apply this concept to carbon dioxide (CO_2) and other greenhouse gases as they rethink regulatory approaches to electric utilities and other major industries. Indeed, in interviews conducted as part of the research for this book, this legislation was repeatedly depicted as a successful model that has influenced policymakers' thinking about the development of greenhouse gas reduction policy.

Third, the Intermodal Surface Transportation Efficiency Act of 1991 offered states new funding for mass transit but also encouraged them to pursue long-term planning to devise effective transportation options that would reduce pollution and energy consumption.[8] This program was intended to prod states to move beyond traditional efforts to maximize new highway construction and consider a range of transportation options. Fourth, the Bush administration created a State and Local Climate Change Program in 1990, which provided a series of grants and technical assistance to aid states in developing the capacity to examine and address climate change. Program grants provided many states with an initial forum for gathering essential data and beginning a process of policy development that made particular sense for individual states.

Then came Rio. Growing international concern about global climate change led to negotiations for a United Nations Framework Convention on Climate Change. Scheduling of the Earth Summit in Rio de Janeiro in June 1992 provided a deadline for negotiations on climate change and a host of other international environmental issues. American support for the Framework Convention was withheld until the Bush administration's demands for some added flexibility had been met, but the president ultimately pledged American support for the convention. In fact, the United States became one of the first nations to ratify the Framework Convention, which was ratified by 167 nations and the European Community within five years.

Like many international pronouncements, the Framework Convention reflected a series of compromises between nations and regions with perspectives that were often wildly divergent. As the energy policy analyst Michael Grubb and his colleagues have noted, "The gulf between negotiators was enormous."[9] Nonetheless, the emerging framework was more than a mere set of platitudes. It provided strong international affirmation, including the voice of the United States, that global climate change is a serious problem whose solution will ultimately require far-reaching cooperation. More important, it established a plan to stabilize greenhouse gases at 1990 levels by 2000. This reflected, according to article 2, "the ultimate objective of this Convention," which was the "stabilization of greenhouse gas concentrations in the atmosphere at a level that would prevent dangerous anthropogenic interference with the climate system."[10] For many nations, including the United States, such stabilization would require significant policy interventions, given anticipated emissions growth during the coming decade. The convention established a mechanism for annual reporting of greenhouse gas emissions by participating nations to help determine compliance with stabilization goals. It also outlined a process for taking additional steps that would add greater specificity to this overall effort and lead to subsequent rounds of policy that were expected to achieve greater levels of reduction.

These modest achievements seemed to set the stage for a major intensification of federal efforts concerning climate change after the 1992 presidential election. George H. W. Bush had campaigned four years earlier as someone who, if elected, would apply the "White House effect" to the "greenhouse effect." As president, he did manage to set in place some basic policy infrastructure. During the 1992 presidential campaign, however, Democrats Bill Clinton and Al Gore skewered many components of the Bush environmental record, including the president's alleged timidity on global climate change. They pledged to abide by the Framework Convention and maintain 2000 emissions at 1990 levels. They also indicated that they would go much further, if elected, and bring new, creative approaches to this problem. Gore assumed a particularly prominent role in these discussions, reflecting his long-standing concern about global climate change and the relatively recent publication of his best-selling book, *Earth in the Balance*. Gore was particularly harsh in his book on those who would adopt incremental strategies to threatening issues such as global climate change: "Modest shifts in policy, marginal adjustments in ongoing programs, moderate improvements in laws and regulations, rhetoric offered in lieu of genuine change—these are all forms of appeasement, designed to satisfy the public's desire to believe that sacrifice, struggle, and a wrenching transformation of society will not be necessary."[11]

Through eight years in office, however, the Clinton-Gore administration experienced considerable difficulty in achieving even "modest shifts in policy" related to reduction of greenhouse gases. It presided over an economic—and greenhouse gas emission—expansion that made a mockery of Framework Convention goals. Between 1990 and 2000, the United States added more than 30 million people and 25 million motor vehicles, tantamount, in terms of energy use, to annexing another state of California. Despite energy efficiency gains in certain sectors, Americans in 2000 used approximately 10 percent more energy per capita than they had in 1980.[12] By 2000, American greenhouse gas emissions had not been stabilized at 1990 levels but had actually increased by nearly 15 percent during the decade. This made the reductions pledged by the American delegation at the Kyoto negotiations—7 percent below 1990 levels by 2012—even more daunting.

The first glimmer of the difficulties the new administration would experience in reducing greenhouse gases occurred during its first year, 1993. Despite strong Democratic majorities in both houses of Congress, Clinton's fiscal reform proposal passed narrowly and survived only after the administration agreed to strip Gore's pet proposal for a tax on the use of fossil fuels. This proposal was never expressly characterized as a greenhouse gas reduction strategy but was clearly intended to begin a process of reducing fossil fuel use—alongside its primary goal of generating additional revenue to help trim federal budget deficits. The political response to this proposal was intense and cut across party lines, leading Clinton to abandon it in attempting to salvage his fiscal program. Clearly, the Clinton administration would have to try something else to achieve greenhouse gas reduction, although its remaining time with the Ninety-third Congress through January 1995 was dominated by the failed effort to develop a comprehensive medical care policy.

After the election of a Republican Congress in November 1994 and a protracted period of interbranch warfare, climate change policy moved even further to the recesses. Each year, the administration advanced some package of incentives to develop clean energy technologies and voluntary programs to attempt to reduce greenhouse gases, along with proposals to increase funding for climate change research. Perhaps the most prominent of these was the Partnership for a New Generation of Vehicles, whereby the administration proposed a far-reaching alliance with the domestic motor vehicle industry to bring a revolutionary set of low-emission vehicles into the marketplace. The program called on each of the Big Three auto manufacturers to develop family-friendly vehicles capable of driving eighty miles on a gallon of gasoline, with "production prototypes" due by 2004. A number of these types of proposals expanded on existing efforts launched in the prior administration, often recommending significant funding increases. Many Clinton proposals met a hostile reception in Congress. Some were approved only after they were amended with a series of formal constraints—including legislative riders named after their chief sponsor, Representative Joseph Knollenberg (R-Mich.)—on federal expenditures or regulations that might be used to attempt to reduce greenhouse gases in accordance with the Kyoto Protocol. In its final years, the Clinton administration's proposals became somewhat bolder, in some instances calling for national adoption of policies that were being developed at the state level, including required levels of energy from renewable sources. These never received a serious hearing, however, and were never aggressively promoted by the president or vice president. Ironically, President Clinton's boldest statement on climate change, a vigorous Internet address in which he endorsed several new proposals, including a "multipollutant" program that included caps on carbon dioxide, took place four days after the November 2000 election.[13]

Of course, the biggest story related to climate change during the Clinton-Gore administration was its engagement in international diplomacy, leading to the signing of the Kyoto Protocol in December 1997. In 1995 the administration had signaled its commitment to an agreement that would move well beyond the Framework Convention when it endorsed the so-called Berlin Mandates that called for development of binding emission reduction targets, exempted developing nations from formal engagement in the process, and set up further rounds of negotiation. The Berlin

Mandates triggered strong concerns in Congress, so much so that the Senate passed a 95-0 resolution in July 1997, called Byrd-Hagel after its chief sponsors, West Virginia Democrat Robert Byrd and Nebraska Republican Charles Hagel. This resolution made clear that the Senate would not ratify any future climate change agreement without commitments from developing countries.

The administration was clearly divided as the Kyoto negotiations approached. Support for active engagement came from the leadership of the Environmental Protection Agency and the Department of State but was tempered by serious reservations from the Departments of Energy, Treasury, Defense, and Commerce. As negotiations sputtered in Japan, Gore reversed his long-standing plan to refrain from direct intervention with a dramatic flight to Kyoto and appearance at the negotiating table, at which he announced that President Clinton had endorsed new flexibility in the American bargaining position. This helped break the logjam and led to an eleventh-hour agreement. As the journalist John Cushman later noted, "After delegates caucused, argued, slumped and snored through the night, toward lunchtime, a pact was reached when the session's chairman decreed that an important conflict would be dealt with later."[14]

Kyoto did indeed leave much to be resolved in subsequent rounds of negotiation, which have continued to unfold in recent years. But the Clinton administration endorsement, which involved a formal signing in New York in November 1998, met with an icy reception on Capitol Hill that cut across partisan lines in both the Senate and House. Opponents contended that the pledged emission reductions were too steep and that the Byrd-Hagel warning on engagement of developing nations had gone ignored. "This is not going to pass the Senate—it's not going to come close," said Hagel, after the United States signed the protocol. "Obviously the president knows that. He's doing something very dishonest by signing the treaty and telling America it's good for them but not having the courage to debate it and try to get a vote on it."[15] Vote counts in the Senate suggested that the administration could rely, at best, on ten to fifteen votes in support, far short of the sixty-seven needed for ratification. Consequently, the administration promised to work on securing greater flexibility in ongoing international negotiations and ultimately convince the Senate of the merits of the treaty. However, Clinton and his administration left office in January 2001 having neither submitted Kyoto to the Senate for consideration nor launched a serious effort to persuade legislators or the citizenry of the merits of the treaty.

A Gore presidency would undoubtedly have led to some revitalized effort to pursue ratification. However, the vice president did not use the 2000 campaign to outline his strategy, and, of course, the divisive election resulted in the ascension of George W. Bush to the White House.[16] Like its predecessor, the Bush cabinet was clearly divided in the early months on how to address global climate change. The EPA administrator, Christine Whitman, and a few allies, including Health and Human Services Secretary (and former Wisconsin governor) Tommy Thompson and Treasury Secretary Paul O'Neill, supported a range of policy proposals to reduce greenhouse gases and continued engagement in international negotiations. Some of the policy options under consideration were in fact initiatives that one or more states had already enacted, including Whitman's New Jersey and Thompson's Wisconsin as well as the president's home state of Texas. But Bush faced considerable opposition from other quarters of his administration and key interest group allies, and in March 2001 he decided to disengage from future discussions related to Kyoto.

In announcing his plan to remove the United States from further international negotiations, Bush vowed to substitute a new domestic plan to reduce greenhouse gas releases. It took nearly a year, however, for him to introduce an initiative that offered a proposal to concentrate not on overall emissions but rather on the "carbon intensity" of the economy. This proposal has been through several iterations, but in essence, the Bush plan asks generators of carbon dioxide to reduce their emissions increase to one-third the rate of economic growth, thereby tying voluntary emission reductions to economic output. A wide chorus of observers has concluded that this is an essentially meaningless goal, because the amount of greenhouse gases generated for each unit of economic activity has declined steadily since the mid-nineteenth century. In fact, by Bush's proposed metric, American "carbon intensity" has decreased by approximately 50 percent since 1950 and is expected to continue to decline without any change in federal or subnational policy.

In making this proposal, Bush did endorse some minor modification of voluntary and incentive programs enacted or proposed by his father and President Clinton.[17] The Department of Energy and its secretary, Spencer Abraham, took the lead role in this effort, under the banner of Climate VISION (Voluntary Innovative Sector Initiatives: Opportunities Now). But President Bush expressly excluded carbon dioxide from a related proposal to reduce conventional air pollutants, removing it from consideration under a multipollutant umbrella that has continued to move forward in a number of states. Relatedly, a number of components of the Bush energy plan, including proposals to expand the use of coal in electricity generation, to intensify exploration for domestic sources of oil, and to maintain standards for motor vehicle fuel efficiency near current levels, would quite likely increase reliance on fossil fuels. Consequently, the Bush plan largely reverses the direction of the first Bush administration and minimizes the likelihood of any serious federal effort to reduce greenhouse gases in the near future.

The American climate change odyssey of the past three presidential administrations has had a significant impact on states and their attempts to define a role in this area. Although their involvement should not be viewed as purely reactive to federal action—or inaction—the very possibility of involvement was influenced by signals from Washington. During the first Bush administration, a number of states did take initial steps, beginning to study the issue and even formulating early policies. States also clearly responded to the opportunities for engagement created by new federal laws on air pollution, energy, and transportation, all of which established either new tools or incentives relevant to development of a long-term strategy for greenhouse gas reduction.

During the Clinton administration, states stepped back somewhat, in part to expand their analytic efforts but also anticipating that the federal government was going to establish some national program, possibly in conjunction with an international climate change regime. After it became evident in 1998 and 1999 that Kyoto was unlikely to be ratified, however, states jumped on the issue with new intensity. That level of involvement has only expanded further in the administration of George W. Bush, when it became clear that the United States had disengaged from international negotiations and that the federal government was unlikely to formulate any serious national strategy to reduce greenhouse gases.

Consequently, a policy problem that had almost universally been defined as a responsibility of international and national governments devolved on a de facto basis in the United States to subnational units. But the disincentives to serious engagement would seem more overwhelming in the individual states than in Washington. If the vast majority of congressional Democrats and Republicans have shown no enthusiasm for greenhouse gas reductions over a decade-long period, why should states be any different? What incentives would states have to unilaterally reduce their emissions, when these would only be a fraction of the national total and would register at best a limited impact internationally? When initiatives like Kyoto were being widely denounced across party lines as a threat to long-term economic sustainability, why would a state enact measures to achieve significant reductions? And if a policy entrepreneur like Al Gore, who rose to national prominence and the Democratic ticket in part on the basis of his passionate commitment to confronting global warming, accomplished little on the issue during two terms as vice president and chose to play down the issue during his presidential campaign, why would a state-based policy entrepreneur even consider engaging the issue?

Shifting to the Statehouse

The very idea of decentralized approaches to combat global climate change is not, of course, an exclusively American conception. Some of the existing international agreements and pronouncements that address climate change endorse—albeit in opaque terms—a significant role for "subnational" governments in policy development and implementation. For example, the Agenda 21 Principles, a forty-chapter action plan for sustainable development adopted at the 1992 United Nations Conference on Environment and Development, provides strong endorsement of a "bottom-up" approach to greenhouse gas policy development. The plan supports substantial experimentation and mechanisms for providing "decentralized feedback to national policies."[18] In turn, article 10 of the Rio Declaration on Environment and Development declares that global warming is most likely to be addressed effectively through broad political participation "at the lowest, most accessible, and policy-relevant" level.[19] As the policy analysts David Feldman and Catherine Wilt have noted, these international agreements "clearly reject centralized, bureaucratic approaches."[20]

These statements reflect a growing consensus that endorses decentralization in many spheres of public policy. The conventional wisdom of the 1960s and 1970s reflected severe doubts about the capacity of state and local governments to protect the environment.[21] The inherently cross-boundary nature of many environmental problems and the potential for localized units to shirk responsibility were seen as insurmountable. Applied to climate change, this conventional wisdom anticipated that states would be highly unlikely to act without exact marching orders—and funds for implementation—from the federal government.

More recent analysis, however, has embraced certain state, local, and regional governments as highly committed to environmental protection.

A growing body of scholarship concludes that these decentralized units are increasingly proving more capable and innovative than their central-level counterparts. Indeed, in some areas of environmental policy analysis, such as the growing body of scholarship on the protection of "common-pool resources" and the emergence of "civic environmentalism," subnational units are regularly depicted as capable of doing little that is wrong, whereas their national counterparts are characterized as doing little that is effective.[22]

The past decade has featured a striking convergence of opinion in this regard. A mounting series of studies and reports from such think tanks as the National Academy of Public Administration, the Brookings Institution, the American Enterprise Institute, and the Hoover Institution, among others, confirms the promise of decentralization. Many endorse far-reaching delegation of decisionmaking authority to state governments, allowing them to play a major role in addressing the "next generation" of environmental challenges.[23]

Climate change has not figured prominently in these studies, and yet many policy areas with direct relevance to greenhouse gas emissions now fall under the purview of state governments. States already implement many federal environmental laws; they issue more than 90 percent of all environmental permits and conduct more than 75 percent of all environmental enforcement actions.[24] They play increasingly central roles in the implementation of clean air policy, ranging from involvement in emissions-trading programs for sulfur dioxide to multistate negotiations over control of ozone pollution. Even air emissions from motor vehicles, long assumed to be an obvious candidate for centralized regulation, continue to be heavily influenced by regulatory initiatives from California and northeastern states.

Similarly in energy, states are increasingly influential players in shaping the essential rules that guide the generation and distribution of electricity. Traditional powers of state public utility commissions have only expanded in recent years, owing to restructuring and consideration of related issues such as reliability and environmental impact in an era of deregulation. Other sectors that generate substantial greenhouse gases, including transportation and waste management, or that could offset global warming through carbon sequestration, including agriculture and forestry, retain substantial roles for state governments. Moreover, states have been in the forefront of policy innovations that could directly influence greenhouse gas releases or be adapted to climate change policy, including pollution prevention, information disclosure, technology sharing, cross-boundary collaboration, and even constructive engagement on common problems with Canadian provinces and other national governments.

Some federal programs may actually serve to increase state engagement on these issues, as did the air pollution, energy, and transportation legislation enacted in the early 1990s. In addition, the federal government has experimented with forms of devolution or power sharing through various administrative agreements. Perhaps the most notable initiatives, and ones with potential ramifications for greenhouse gases, were a series of Clinton administration experiments that gave more latitude to state officials in exchange for demonstrable improvements in environmental performance. In some instances, this involved negotiated settlements for specific facilities, such as Project XL, which was part of the administration's strategy to "reinvent government." This initiative, however, included a limited number of facilities and revealed considerable difficulties in securing multibranch cooperation.[25] A potentially more far-reaching tool is EPA's National Environmental Performance Partnership System (NEPPS), which offers states the prospect of increased flexibility in regulatory interpretation, policy prioritization, and federal grant use in exchange for formal commitments to pursue innovation and achieve measurable improvements in environmental performance. More than forty states have formally participated in NEPPS to some degree, although the program's actual impact on fostering policy innovation appears quite limited in most states to date.[26]

The Rapid Evolution of the State Role in Climate Change

It is possible to identify distinct periods in state approaches to climate change over the past decade. Many of the most prominent state climate change initiatives are of fairly recent vintage, having been approved after the negotiation of the Kyoto Protocol in December 1997. A number of these remain in the earliest stages of policy implementation. Yet at least some states were clearly thinking about this issue and taking some early steps nearly a decade before Kyoto.

This initial stage of innovation coincided with the ascendance of climate change in the media and at the national level in the late 1980s. Many early state programs focused on chlorofluorocarbons (CFCs), reflecting growing concern about their impact on both depletion of the ozone layer and global warming. These efforts were ultimately eclipsed by national and international agreements on CFC elimination, but they helped put climate change on the agenda of many state governments. Despite the early emphasis on CFCs, a small subset of states passed legislation or executive orders between 1988 and 1990 that expressed concern about other greenhouse gases and endorsed some initial steps.[27] Many of these were focused on the activities of state governments, commonly promoting greater energy efficiency in government-operated buildings and vehicles. These were similar in many ways to what President Clinton proposed for federal institutions a decade later. New Jersey Executive Order 219, for example, signed by Republican governor Thomas Kean in February 1989, called upon all units of state government to take the lead in reducing greenhouse gases. This and related initiatives were largely symbolic, setting nonenforceable recommendations and lacking any resources for implementation. They continue, however, to be recognized in some state policy circles as having established a precedent for further state action.

After this initial flurry of activity, many states pursued analytical work on their greenhouse gas emissions and began to review future policy options in the first half of the 1990s. Federal grants from EPA's State and Local Climate Change Program were combined with state resources to underwrite detailed reviews of greenhouse gas sources and emission trends within individual states. Dozens of states produced detailed "greenhouse gas inventories"; some used these to formulate "action plans" outlining various strategies for emission reduction. These analyses provided an empirical foundation for much subsequent state policy activity and also served, in many states, as an initial opportunity to bring together constituents from diverse state agencies, industries, universities, and advocacy groups to meet and consider climate change as a state policy issue. Multistate networking also began during this period, often through conferences or research reports that allowed states to begin to think more collectively about these issues. All of this activity occurred alongside initial state efforts to respond to new federal policy established on air pollution, energy, and transportation in the early 1990s, which further encouraged states to begin to consider greenhouse gases and utilize new policy tools—such as emissions trading and integrated energy planning—that might prove applicable to larger climate change strategies in the future.

Such analysis continues in many states, often under the auspices of a state agency unit given lead designation for climate change work. Indeed, virtually every state with an active climate change program examined in this book has at least one official who is widely perceived as a leading expert on the subject, both in how it might affect a particular state and in the development of various greenhouse gas reduction policies. In these states, such professionals often carry the informal title of "Mr. Climate Change" or "Ms. Climate Change" for state government. They are pivotal players in moving the policy process forward.

All of this work set the stage for a much more active period of policy formation in the late 1990s. Some significant state laws were approved during this period, ranging from Minnesota's 1993 legislation to include the environmental and economic impacts of carbon dioxide releases as a formal component of decisions on energy development to Oregon's 1997 law that established carbon dioxide emissions standards for any new electrical power plants opened in the state. At the same time, a significant number of states moved in a different direction, reflecting the burgeoning controversy surrounding Kyoto. During 1998 and 1999, sixteen states (Alabama, Arizona, Colorado, Idaho, Illinois, Indiana, Kentucky, Michigan, Mississippi, North Dakota, Ohio, Pennsylvania, South Carolina, Virginia, West Virginia, and Wyoming) passed legislation or resolutions that were highly critical of the Kyoto Protocol and opposed ratification by the U.S. Senate. Many of these were purely advisory, employing near-identical language from state to state. Some states, however, chose to go further and prohibited any unilateral steps to reduce greenhouse gases. In West Virginia, for example, legislation passed in 1998 prevented state agencies from entering into any agreement with any federal agencies intended to reduce the state's greenhouse gas emissions.

State efforts to contain involvement on climate change have been eclipsed in more recent years with an unprecedented period of activity and innovation. New legislation and executive orders intended to reduce greenhouse gases have been approved in more than one-third of the states since January 2000. Multiple programs have been enacted in some states, and many new legislative proposals are being advanced in a large number of states. These new programs include formal carbon dioxide caps on particular industries, statewide goals for greenhouse gas reductions, formal agreements with utilities and industries to reduce carbon dioxide emissions, mandates to generate specified levels of electricity from sources that generate no greenhouse gases, mandatory reporting of carbon dioxide emissions, voluntary registries for industries seeking credit for reductions in any future regulatory regime, and de facto "carbon taxes" on utility bills that create pools of funds for energy efficiency, among others.

More recently, California jumped into the fray in July 2002 by initiating a decade-long process for establishing carbon dioxide emission standards for vehicles, building on other states' efforts to reduce greenhouse gas releases from transportation sources.

Some states have even begun to work collectively, perhaps best reflected in the efforts of the six New England states to establish regionwide standards and programs in concert with Quebec and the four Maritime Provinces of Canada. This initiative may be expanded in coming years to include additional states and provinces. Despite the common goal of greenhouse gas reduction, there is enormous state-by-state variation in these programs and little sign of formal diffusion of particular programs or policy ideas, aside from the clear diffusion pattern in the anti-Kyoto enactments. However, this pattern may change as the policy area matures, involving a larger number of policies with overlapping components.[28] For example, Nebraska's passage of a program designed to sequester carbon through agricultural practices was signed into law in April 2000 and prompted Illinois, North Dakota, Oklahoma, and Wyoming to pass legislation with virtually identical language within one year. In turn, some states give indication of formally joining forces in an attempt to prod the federal government to take more significant action. For example, the attorneys general of Connecticut, Maine, and Vermont filed suit in federal court in June 2003 in an attempt to force the Bush administration to address carbon dioxide as a pollutant under the Clean Air Act.

State Differentiation and the Role of Policy Entrepreneurs

States have clearly picked up the pace in climate change policy, as measured by the number of new initiatives, their range of coverage, and the rigor with which they attempt to achieve significant reductions. What remains less obvious, however, is why so many states have found it increasingly possible politically to enact policy that continues to defy any consensus at the federal level and for which there is no international or national mandate to take action. Why, in turn, is there such enormous state-by-state variation in response? Why may states that are comparable in geography, likely impact from climate change, level of membership in leading environmental groups, and even partisan control of state government institutions take dramatically different approaches to the same issue?

The recent proliferation of policy innovation demonstrates the variety of policy options open to states, as the diverse range of case studies introduced in subsequent chapters suggests. Conventional analysis of environmental issues might anticipate that factors common to formation of a great deal of environmental policy at the national or international level might also be evident in these state cases. In particular, one might expect that key states may have experienced localized disasters that could be directly linked to climate change and thereby galvanize public opinion. Such "focusing events" might have been seized upon by major environmental groups, which worked closely with sympathetic legislators and governors and journalists to push aggressively for new policy in the face of intense opposition from parties that would incur the costs of regulatory compliance. This might lead to the development of new and formidable advocacy coalitions, which would attempt to garner broad public support for efforts to respond to these serious environmental threats. Perhaps relatively recent outbreaks of drought in agricultural states, coastal calamities in states with lengthy ocean borders, and violent weather outbreaks such as floods and tornadoes created a groundswell of public concern—and a window of opportunity—that forced states to respond to climate change.

But this is not what occurred in the states examined in this study. Instead, a much quieter process of policy formation has emerged, even during more recent years, when the pace of innovation has accelerated and the intent of many policies has been more far-reaching. This is not to suggest that climate-related episodes have been irrelevant or that leading environmental groups have played no role in state policy development. Contrary to the kinds of political brawls so common in debates about climate change policy at national and international venues, however, state-based policymaking has been far less visible and contentious, often cutting across traditional partisan and interest group fissures. It has, moreover, been far more productive in terms of generating actual policies with the potential to reduce greenhouse gas releases.

In many respects, states have had a relatively quiet decade in which to think about climate change, in terms of both how serious an environmental problem they perceive it to be and how they might fashion their own policies to reduce greenhouse gases. In some instances, states are clearly responding to a perception that climate change is real and that there is a serious need to craft policies as soon as possible. But these responses are also coupled with efforts to design policy that "fits" the economic and political realities of a particular state. These are intended to minimize any economic disruptions that might occur during implementation and to take maximum advantage of economic development opportunities that may stem from early action on climate change. These opportunities may include opening new markets for emissions reduction technologies or generating "credits" through reduced emissions that can ultimately be sold in national or international markets. They may also include compacts with regulated industries that offer new flexibility in regulation or guarantees that any early greenhouse gas reductions will receive full credit in the event that national and international institutions get their respective acts together and establish formal rules for climate change policy.

What has been missing in these state policy processes is the kind of anguished, often moralistic, rhetoric that has polarized national debate and made any semblance of consensus at that level so elusive. Instead, state policy deliberations over climate change may have benefited from a kind of political cover provided by the widely held presumption that states had neither incentives nor resources to play any serious role. Many national interest groups—environmental and industrial—as well as the media and scholarly communities have essentially ignored what states were doing and instead assumed that the real action was occurring in Washington, D.C., or the various cities—from Kyoto to Bonn—where periodic rounds of international diplomacy were being played out.

Many states used this extended period to reflect seriously about the issue of climate change and how they might respond to it. Many built on early symbolic initiatives and used federal resources to study the climate change issue and consider how it might affect them. They also began to develop formal ways to measure their greenhouse gas releases and to consider reasonable steps that might allow them to reduce these releases. At various points, these efforts took institutional form, such as creation of a cross-agency task force or working group. This often included the establishment of a unit given lead responsibility for ongoing analytical work and policy development. Such units were often inserted into existing state agencies, such as those involving environmental protection, energy, or natural resources, usually small in number of staff but diverse in disciplinary and technical expertise.

These units provided a base, in many states, from which agency officials could evolve into prominent "policy entrepreneurs"—individuals who command widespread respect for their expertise on a given issue and their integrity as credible brokers of information. They are often well positioned to see opportunities for new policy and to literally translate ideas for innovation into workable policies. These kinds of entrepreneurs are, of course, far more than idea generators. They often assume a central role in building and sustaining the coalitions that will be essential in transforming an idea into a policy—such as legislation or an executive order—that reflects a broad base of support. This coalition building includes effectively making the case for a particular policy within agency circles, often convincing an agency commissioner or secretary of the merits of a new proposal. This is a crucial beginning in the process of securing a wider base of support, sustaining resources for the innovating unit, and deflecting opposition from other governmental units that might prefer to see the state pursue other policies that better reflect their own particular interests.

Entrepreneurs must also know how to play outside-agency politics, finding allies who will be attracted to the study of climate change and consideration of policy development. In many states, affiliates of large environmental groups, industry groups, or labor unions have been conspicuously absent from this policy development. "We are busy fighting the tanks on the battlefield that are shooting at us right now," explains a senior member of a state affiliate of one of the nation's largest environmental groups, in a representative comment. "And there are a lot of issues out there with greater immediate visibility and perceived threat to our members than climate change." With many traditional players otherwise occupied, entrepreneurs search for opportunities through alliance with less visible partners, such as smaller environmental groups with particular expertise in emissions trading or a subset of industries prepared to consider significant greenhouse gas reductions in exchange for certain benefits the state might grant. In turn, entrepreneurs must begin to build bridges, with the help of their allies, to legislators and governors—and their aides—who can be persuaded of the political benefits of supporting such steps.

In many instances, these entrepreneurs operate well below the public radar screen. Most receive little or no public recognition for their efforts, even at the moment a policy is approved by a legislature or governor. They tend not even to get their names in the state newspaper of record when policies are enacted or prominent seats at ceremonies when these policies are signed into law; any claiming of credit is dominated by elected officials and other more prominent individuals. Many even lack financial support, which constrains their travel to conferences or to meet with colleagues from other states or the regional offices of federal agencies.

In most cases, these entrepreneurs function within what the political scientist Daniel Carpenter has described as the "mezzo level" of public bureaucracies. This level falls below the "executive" tier, which consists of cabinet-level appointees who tend to report directly to the chief elected official such as the governor. But it is clearly above the "operations level," which is dominated by officials absorbed with the details of policy implementation, often at the street level. As Carpenter notes in his study of federal-level agencies, officials at the mezzo level tend to have sufficient expertise and latitude to be a source of policy development. They often work effectively to build essential coalitions of networks, including key executive-level allies such as department heads. In some instances, they leap at an opportunity when an executive-level official indicates a willingness to "do something" about a given issue. "The hierarchical structure of many bureaucracies," according to Carpenter, "leaves middle-level bureaucrats in the best position to experiment, learn, and innovate."[29]

In many of these state cases of innovation in climate change policy, mezzo-level entrepreneurs play a pivotal role. These officials consider what they are doing to be among the most exciting and meaningful work of their careers. They clearly hope that their efforts will allow them to remain active in this area. Most have not been groomed for careers in state climate change analysis and policy; many acknowledge that they never even studied the issue in college or graduate school. But they have been prepared for their roles by prior expertise in related areas of policy. In fact, leading state policy entrepreneurs draw heavily on ideas and experience gained in different but highly relevant realms. These include work on emissions-trading programs to combat conventional air pollutants, involvement in prior generations of policy to promote greater energy efficiency and development of renewable energy sources, and engagement in policies to promote recycling of solid waste. Each of these areas offers a type of "path dependence," providing ideas and insights that entrepreneurs translate and apply to the issues involved in addressing climate change.

The very notion, of course, that agency-based entrepreneurs have assumed lead roles in the formation of climate change policy in a diverse set of state capitals may seem counterintuitive. An entrepreneurially based approach to the development of climate change policy challenges much conventional analysis of agency officials and their behavior. A great amount of academic and journalistic ink has been spilled in recent decades accounting the dysfunctional qualities of such officials, more commonly known as bureaucrats. These analyses take different forms but often question the basic competence of these officials and their ability to function in an effective manner. One important line of scholarly work notes a growing pattern whereby elected officials at federal and state levels—commonly known as "principals"—have increasingly attempted to constrain the behavior of bureaucrats—widely characterized as "agents." Under this approach, elected officials are clearly in command. They compose increasingly detailed legislation and rules to maximize the likelihood that agency officials will be confined to carrying out exact marching orders and face punishment for efforts to "shirk" their duties or "sabotage" these commands.

This study's emphasis on entrepreneurs as central agents of change is not intended to suggest that bureaucrats have somehow broken the chains of the principal-agency bond and are running wild. Larger political and economic contexts can influence whether agency officials can legally even talk about climate change policy, much less take steps to develop it. In particular, states with economic interests threatened by any unilateral state action to reduce greenhouse gases may quash any serious opportunities for entrepreneurial activity. Nor is this analysis intended to suggest that these state cases reflect an American version of the popular British situation comedy, Yes, Minister, in which bureaucrats regularly outsmart clueless elected officials and thereby call the shots.

But it does indicate that at least some officials in some states possess far greater analytical and political skills than are generally associated with bureaucratic behavior. Indeed, this finding may be consistent with a growing response to much conventional analysis. As the political scientists John Brehm and Scott Gates have noted, "Despite significant efforts to constrain bureaucratic choices, bureaucrats possess significant degrees of discretion."[30] In all the states in this study that have pursued climate change policy initiatives, state agency officials have retained such discretion and utilized it to assume central roles as policy entrepreneurs. This finding may be consistent with a growing literature on policy entrepreneurship that explores the roles that agency officials—and other entrepreneurs—may play in both the domestic and international arenas.[31]

In the case of state policy to reduce greenhouse gases, policy entrepreneurs may have taken advantage of three distinct elements of this evolving issue area. First, the decade-long failure of federal government institutions to enact even the basic design of an emissions reduction policy has clearly created "policy room" for state experimentation. Much scholarly and media analysis of climate change has presumed that some grand, internationally recognized regime would soon be established and thereby impose the rules of the game on all participating jurisdictions.

State-based entrepreneurs have proven increasingly skeptical of federal action, as well as the long-term viability of Kyoto, in making the case for states to take the lead, both to achieve early reductions that may receive credit in the longer term and to influence the shape of future federal and international policy.

Recent experience would suggest that climate change is such a complex, cross-cutting issue that there is not "one best way" to reduce greenhouse gas emissions or even a single economic or policy sector upon which reduction can best be concentrated. Even international agreements with some measure of support outside the United States, most notably Kyoto, are riddled with loopholes and implementation uncertainties among participants.[32] Different states may be unusually well equipped to fashion reduction strategies that make sense, given their particular mix of economic and governance realities and the fact that no government or private entity has mastered "how to do" climate change policy. This uncertainty has created a national policy void that has proved increasingly inviting to state policy entrepreneurs.

Second, state-level policymaking is often quite different from what occurs in Washington. As at the federal level, state governments can bog down in partisan squabbles and succumb to the powers of influential interest groups. But in many states, policymaking is far more informal, and entrepreneurial opportunities may be considerably greater, than in Washington. In the absence of particularly strong opposition from interest groups, entrepreneurs may have a much better opportunity to establish and sustain supportive networks. These may involve other agencies, interest groups, or allied elected officials and may have been established over an extended period, over a decade in some of the state climate change cases. The mezzo level in many state agencies, such as environmental protection and energy, is much less densely staffed than in their federal counterparts, and the layers between an agency and the governor's office are likely to be much thinner. This allows an individual to emerge as the trusted resident expert on a particular topic, such as climate change, able to get important messages to prominent places in the state governance structure. Consequently, many state capitals may offer particularly promising entrepreneurship opportunities, particularly for relatively "new" issues for which an infrastructure of established policies and interest group positions has not been created.

Third, and perhaps most significant, states may be increasingly inclined to take leading-edge action on environmental issues, even reduction of greenhouse gases, because they perceive it to be in their economic self-interest to do so. Such a stance reflects a generation of evolution in state economic development and environmental protection policy. As a result, an earlier tendency to depict environmental protection efforts as posing a zero-sum trade-off with economic growth has long since been supplanted in many states by a more nuanced view that recognizes a variety of economic advantages from environmental protection.

State governments remain remarkably sensitive to economic development.[33] Governors and legislators are acutely aware that their long-term fortunes may hinge on electorate perception of their economic stewardship. This explains the massive efforts made by most states to respond to the needs of established firms at the same time that they are constantly searching for ways to underwrite future development opportunities, whether home grown or lured across state boundaries. Conventional thinking about state politics has presumed indifference to environmental protection, particularly in instances in which at least some locally generated contamination might migrate across state or national borders. States, in short, have been deemed so determined to sustain local economic activity that they would not want to take any steps to unilaterally impose regulations and thereby increase the cost of doing business within their boundaries. Such an approach has fueled long-standing opposition to more decentralized regulatory strategies, assuming that any deviation from federal command-and-control policies would unleash a "race to the bottom." In such a race, states would presumably trip over one another in trying to set the lowest possible standards and thereby appease in-state firms eager to minimize compliance costs.

Such a perception seems decreasingly compelling when weighed against the preponderance of evidence concerning state government innovation and expansion of capacity to address environmental problems. One clear motivating factor involves quality-of-life calculations, whereby states seek to project environmentally clean images and maximize their attractiveness to potential residents and investors. This may be especially evident in the case of states long derided for having environmental quality that is so poor as directly to impair economic development efforts. New Jersey, which figures prominently later in this book, may be the premier example of this phenomenon.

Greenhouse gas reduction may appear, at first glance, a very long-term strategy for economic development by state governments. But state receptivity to a range of climate change policies suggests that supporters perceive multiple benefits from early action, many of which could offer immediate economic payoffs. In one set of cases, state concern about adverse effects of climate change is very real. Reduction of greenhouse gases has been received as the beginning—albeit small—of a credible response to protect economic concerns, whether oceanfront development or preservation of tourism. In most others, however, supplemental impacts from reducing greenhouse gases may have spillover benefits, including diversification of energy supply, coordinated reduction of overall air emissions, increased agricultural productivity, reduced traffic congestion, and longer-term regulatory predictability for regulated firms, among many others. In short, the conventional analysis of strategies for greenhouse gas reduction that assume the only conceivable benefits stem from any long-term impact to slow the pace of climate change misses the mark when one looks at the more expansive set of factors motivating states to act. In all of the cases presented in this book, economic development has been a contributing—and, in some instances, dominating—factor behind state greenhouse gas reduction policy.

Issue Framing and Policy Labeling

Just as no two state polities are identical, not all states provide conditions whereby entrepreneurs may emerge, much less have an impact on policy. Figure 1-1 outlines a framework for explaining the significant variation demonstrated by respective states in moving into the area of climate change policy. Viewing states through the lens of this framework reflects case study analysis of their past decade of experience—or lack thereof—in various aspects of policy related to climate change. For the kinds of policies addressed in this book, issue framing and policy labeling are particularly important components in understanding the varied terrain upon which potential entrepreneurs must operate at the state level.

Issue framing reflects the most common way in which a policy issue has come to be characterized—or defined—in a given political system. Framing explains the ways in which, according to the political scientist Bryan Jones, attention has been directed "to one attribute in a complex problem space."[34] It thereby reflects the way state political actors and the general citizenry tend to conceptualize an issue, implicitly defining boundaries of what may be possible before the particulars of a given policy are proposed. As the political scientist Shanto Iyengar has noted, "The manner in which a problem of choice is 'framed' is a contextual cue that may profoundly

Figure 1-1. *Framework for Explaining State-by-State Approaches to Greenhouse-Gas Reductions*

Issue framing	Policy labeling	
	Explicit	Implicit
Response to environmental threat	Prime-time strategies —CO_2 and greenhouse gas regulatory standards —Mandatory CO_2 reporting —Statewide greenhouse gas reduction commitments —Industry reduction covenants (New Jersey, Oregon, Wisconsin, New Hampshire, all New England states) <div align="right">1</div>	Stealth components of prime-time strategies —Social benefit charges for energy —Energy efficiency <div align="right">2</div>
Response to economic development opportunity	Opportunistic strategies —Agricultural carbon sequestration —Forestry carbon sequestration —Technology transfer agreements (Illinois, Nebraska) <div align="right">3</div>	Stealth strategies —Renewable portfolio standards —Vehicle use reduction (Texas, Georgia) <div align="right">4</div>
Response to economic threat	Hostile strategies —Bans on any state agency actions to reduce greenhouse gases —Anti-Kyoto resolutions (Michigan, Colorado) <div align="right">5</div>	Indifferent strategies —Disengagement from interstate discussions —Failure to apply for federal funds (Florida, Louisiana) <div align="right">6</div>

influence decision outcomes."[35] Policy entrepreneurs may devote considerable attention to framing an issue in such a way that a policy response becomes compelling.

In contrast, *policy labeling* is more precise. It describes the explicit language used to describe policies that may be attempted, given the opportunities provided (or constraints imposed) by issue framing. Elected officials are highly sensitive to policy labeling, clearly eager to claim credit and avoid blame for their actions. The way in which various policies are labeled may help them maximize credit or minimize blame.[36] Labeling thus becomes an important—and revealing—aspect of the strategic development of climate change policy.

Issue Framing for Climate Change

At the state level, issue framing helps explain how climate change has come to be characterized and whether some policy response may be merited. It also helps guide what kinds of policy responses, if any, may be possible. Framing may help determine whether policy entrepreneurs can literally survive in a state agency, much less successfully influence policy. For environmental issues for which there has been a dramatic focusing event, such as an oil spill or a massive release of toxic chemicals, the issue is likely to be framed as a disaster that merits immediate and extensive action. But not all issues provide such clear framing.[37] In fact, in areas in which focusing events—and strong political advocacy—have been lacking, opportunities for entrepreneurs to contribute directly to issue framing may be significant, whether through release of studies and testimony, development and advocacy of policy proposals, or formation of coalitions.

The cases considered in subsequent chapters suggest that states may frame their policy reactions to climate change in one of three ways: as a response to an *environmental threat*, as a response to an *economic development opportunity*, or as a response to an *economic threat* (see figure 1-1). In some instances, a state may determine that climate change is a serious environmental problem: if ignored, the accumulation of greenhouse gases may pose significant environmental threats to that state. Under these circumstances, a policy response designed to reduce greenhouse gas emissions is warranted. But the state response will be tempered by an attempt to minimize economic disruption and will, to the extent possible, use any intervention available to foster economic development. A number of strategies make such a linkage possible. For example, a state may give special designation to a firm that achieves greenhouse gas reductions and offer it various benefits, such as more flexible compliance schedules for routine inspections or greater latitude in changing product design—and related emissions—as long as overall releases stay below an established cap. In such cases, policy entrepreneurs have maximal latitude both in framing climate change as an issue warranting a serious response and in crafting a significant policy or set of policies.

In another set of cases, states may not view climate change as a major environmental challenge but may instead identify promising opportunities for economic gain by enacting policies that reduce greenhouse gases. These may allow a state to seize opportunities presented by evolving international markets for emissions-credit trading. It might also position a state to become a leader in the development of a particular technology and skill that could eventually be exported to other nations and states. The policy may acknowledge a potentially salutary environmental impact through reduction of greenhouse gases, but this is not a primary motivating force behind its formation. In fact, without the possible economic payoff, it is highly unlikely that the state would take this policy step. Measuring and marketing "carbon credits" through agricultural and forestry practices that actually offset greenhouse gas emissions is one obvious approach that may fit this set of cases. In these cases, entrepreneurs will accentuate the economic development benefits of such a strategy, with any environmental improvement derived from greenhouse gas reduction "framed" as only a supplemental benefit.

In still other cases, a state may react to climate change in a manner similar to the federal government. It may or may not acknowledge climate change as a potential long-term environmental threat but is principally motivated by alarm that any serious response could have severe short-term economic repercussions. Indeed, in these cases, the anticipation of negative economic impacts from any efforts to reduce greenhouse gases clearly outweighs any potential benefit that a state might derive. Such states may develop new policies concerning climate change, but these will be designed to thwart state officials—and potential entrepreneurs—from taking any steps that could result in greenhouse gas reduction. Much as the Knollenberg Amendments have served to constrain potential actions by federal agencies, states that frame climate change principally as an economic threat in the event that any reductions are attempted will work aggressively to prevent that eventuality. In these cases, entrepreneurs may lie low in anticipation of some future shift that might create an opportunity more receptive to greenhouse gas reduction, perhaps through a change in elected officials, agency leadership, or the stance of influential interest groups in the state.

Policy Labeling for Climate Change

Because different states frame possible responses to climate change in different ways, the labels used to describe their policy responses may reveal their strategic responses to these conditions. Entrepreneurs may be particularly well positioned to influence labeling decisions to thereby maximize the likelihood of political appeal. Labeling may be particularly sensitive in this area, given the large amount of scientific and political controversy surrounding the issue of climate change. In fact, sensitivities may be so great that a state may make no effort to acknowledge that its policies might reduce greenhouse gases. Instead, it may select labels that highlight other attributes of the policy and thereby obscure any potential impact it may have on climate change. The policy may, of course, be relabeled at some future point when it becomes politically appropriate to do so, such as a point at which a state—or its governor—might be eager to demonstrate that it is doing something to combat climate change.

Given the ubiquity of greenhouse gases and the wide range of policies that may influence their generation, potential reduction policies do not have to use explicit labels such as carbon dioxide, methane, or greenhouse gases to describe what they intend to do. In some states, policy labeling will be explicit, however the issue is framed. For example, if officials want to claim credit for a response to the environmental challenge of climate change, they may well use specific labels such as "greenhouse gas reduction" or "carbon dioxide mitigation." States eager to seize the economic development advantages of early involvement in climate change policy may similarly be eager to be explicit about their intent. For New Jersey governor Whitman, explicit labeling of most elements of that state's climate change policy afforded a superb opportunity to claim credit. The issue of climate change had been increasingly framed in New Jersey in the late 1990s as a significant environmental threat for which a serious—but economically feasible—policy response was appropriate. Whitman had been widely criticized for budget cuts that resulted in significant reductions in state environmental management staff. She had also been attacked for taking an allegedly weak approach to the enforcement of existing environmental laws, seen as overly deferential to the preferences of regulated parties.[38] Thus an executive order explicitly labeled to address greenhouse gases—issued at a well-attended ceremony—allowed her to claim credit for responding to a new environmental problem and deflect blame for the way she had handled previous ones.

In contrast, other states may choose to play down the greenhouse gas reduction elements of a particular policy and instead accentuate other features. The reduction may well occur, and it may be carefully monitored by the state for potential future use. But the policy will be labeled in such a way as to make it more clearly palatable politically and suitable for the way the issue is framed in a given state. Consequently, officials will be able to claim credit for elements of the policy that are more attractive—such as increased energy efficiency, reduced traffic congestion, long-term regulatory predictability, or improved air quality—and deflect any potential blame for getting too far ahead of the rest of the nation on a controversial issue such as climate change. For example, Texas officials intentionally chose not to label their renewables portfolio standard as an initiative to reduce greenhouse gas emissions, even though all parties recognized that aspect of the proposal and the state has continued to monitor actual reductions. There is no reference to greenhouse gases or carbon dioxide in any part of the 1999 legislation. But as the possibility increased that Governor Bush might become President Bush—and as the program stimulated an unexpectedly huge growth in the supply of renewable energy—a gradual relabeling has occurred whereby Texas officials are increasingly inclined to speak in explicit terms about the greenhouse gas ramifications of the program. Even George W. Bush has begun to use the program as an indicator of his commitment to greenhouse gas reduction, although he never applied such a label in public at the time the bill was signed in 1999.

Differing Opportunities and Constraints

Blending issue framing and policy labeling into a common framework creates six distinct cells that differentiate individual states from one another as well as the kinds of policies they are likely to generate (figure 1-1). Each of these cells is introduced briefly below, including some of the types of policy responses that might be anticipated from each set of states. It is also used in subsequent chapters to more fully differentiate specific state cases and their respective policy responses.

Prime-time states (cell 1) have formally recognized climate change as a serious environmental threat and have responded with multiple programs that explicitly establish greenhouse gas reduction as a state policy goal. Such states may respond with a comprehensive initiative that creates a statewide reduction goal and then delineates strategies for achieving its pledged reductions. Other prime-time states may have approved various elements of a strategy through separate laws and executive orders in moving toward a comprehensive strategy. Prime-time states are those in which governors have "gone public" in their support for greenhouse gas reduction. They have made multiple pronouncements that they take this issue seriously yet want to tailor state response to economic opportunities and concerns. Entrepreneurs have generally been active for much of the past decade in these states, trying to push the state as far as is politically feasible. In this study, New Jersey, Oregon, Wisconsin, and New Hampshire (as well as the entire cluster of New England states) belong in cell 1. Illinois also demonstrates some of the characteristics of this type of state but does not fully fit into this cell.

Prime-time states may, in some instances, choose to play down the specifics of some elements of their strategies. Labeling may come into play in these cases—for example, when a state wants to include greenhouse gas reductions from a particular policy in its overall approach but takes the political decision to delete any specific references to carbon dioxide and greenhouse gases. Such *prime-time stealth* states (cell 2) choose to accentuate other aspects of the proposed policy instead. For example, a number of states place fees on electricity bills, most commonly known as "social-benefit charges," allocating generated funds to energy efficiency or development of renewable energy sources. These clearly have the capacity to reduce greenhouse gases through reduced electricity consumption, but states have been wary of labeling them as either "taxes" or "climate change" policies and instead emphasize other energy-related goals.

In contrast, *opportunistic* states (cell 3) have developed one or more programs to reduce greenhouse gases and are primarily motivated by opportunities for economic development. Such states have not registered particularly strong concerns about climate change as an environmental threat, although they are not hostile to taking steps to reduce greenhouse gases—as long as economic development opportunities are paramount. These states tend to produce policies that pay only passing reference to environmental impacts and instead almost exclusively emphasize the policy's economic objectives. In terms of labeling, they are explicit in outlining ways in which carbon dioxide or greenhouse gas mitigation will be addressed. These types of states tend to have fewer overall greenhouse gas reduction programs, unlike their prime-time counterparts. They may, in fact, have only a single initiative in place. In these cases, governors and prominent elected officials ultimately claim credit for the policy but focus almost exclusively on its linkage with other economic development strategies of the state. Entrepreneurs in opportunistic states tend to be relative newcomers to the climate change issue, compared with their counterparts in prime-time states. Nebraska is a primary example of an opportunistic state, viewing necessary agricultural reforms as a potential "carbon cash crop." Many elements of Illinois's approach to this issue also are consistent with the opportunistic definition.

Stealth states (cell 4) may pursue somewhat similar approaches as opportunistic states and be similarly motivated by economic development opportunities. But these states, as suggested in the Texas case noted above, have made a clear decision to refrain from referring to their policies as related in any way to greenhouse gas reduction. In these instances, state officials are indeed aware of likely reductions and take steps to measure them to ensure credit for use in any future reduction strategies. Hence reductions are best thought of as "incidental" rather than "accidental." But governors and elected officials in these cases make no effort to claim credit for the greenhouse gas reduction efforts and may not want key constituencies to realize they are even considering the issue. Indeed, they may fear that explicit reference to greenhouse gas impacts could serve to block the policy and also mitigate any political benefit they might derive from accentuating anticipated economic benefits. In these cases, entrepreneurs who promote the policy for its greenhouse gas impacts must walk a particularly delicate line; they must be adept with labeling and careful to play down any climate change impacts. "Shhh, don't say the term 'greenhouse gases' too loudly," hushed a prominent policy entrepreneur in a stealth state. "We're getting something done here, on conventional air pollutants *and* greenhouse gases. We sure don't want to blow that." Of course, relabeling can occur, again illustrated in the Texas case, as political conditions change. In this study, Texas and Georgia clearly belong in the stealth category, although some elements of Colorado's approach also fit here.

Finally, some states may not perceive climate change as a credible environmental concern. They may instead view any policy that would unilaterally reduce greenhouse gases as a serious threat to their economic well-being. States taking this stance are likely to be those with one or more prominent industries that generate large quantities of greenhouse gases and would oppose unilateral state reduction efforts. States with large auto-manufacturing capacity (such as Michigan), major coal-mining operations (such as West Virginia), or massive coal-burning utility plants (such as Ohio) would appear least receptive to entrepreneurial actions to reduce greenhouse gases. In fact, such states have been among the most resistant to making any effort to reduce emissions to date. These states may respond in one of two ways. Some will become outwardly *hostile* (cell 5) to any possible intervention and take steps to preclude that possibility. Such states may go so far as to formally prohibit state officials—potential entrepreneurs—from taking any steps to develop or propose policies that would have the effect of reducing greenhouse gases. They may also express formal unwillingness to work cooperatively with federal agencies or even refuse federal funds for research. Such states clearly want nothing to do with climate change policy development, and their elected officials may use such policies to demonstrate to key constituents their determination to protect their interests. In fact, they may boast of their indifference to greenhouse gases and deride any concern over the issue as environmental hysteria. Michigan is a leading example of a hostile state in this study, although Colorado also demonstrates many such characteristics.

A more subtle variation of this approach is taken by *indifferent* states (cell 6), which do not formally go on record with prohibitive policy. Nevertheless, these states also choose to refrain from any involvement in climate change policy development. They secure their goals by not saying or doing much of anything about climate change. They apply belatedly, if at all, for federal grants; they do not attend regional or national meetings on the subject or actively pursue reduction policies; and they implicitly make clear to their state officials that climate change is largely off-limits as a potential area of activity. Florida and Louisiana epitomize this type of state.

Obviously, the roles of potential entrepreneurs will vary markedly by the kinds of political conditions they face. Entrepreneurs in prime-time states may have considerable opportunities to do their thing, whereas counterparts in hostile states may face sanction—or unemployment—if they so much as discuss the issue. In fact, in hostile states, entrepreneurial opportunities related to climate change may clearly be better for those officials eager to win the support of senior appointees and elected officials by demonstrating their commitment to stomp out any contrary activity. In contrast, of course, entrepreneurial opportunities expand the more issue framing moves in the direction of the prime-time cell. In such cases, states may emerge with fairly comprehensive strategies to reduce greenhouse gases. They may also enjoy both a supportive coalition and sufficient managerial competence to implement these programs. Ironically, given federal disengagement of the United States in Kyoto and related international negotiations, some of the prime-time states may be more capable of meeting Kyoto-type reduction goals than many of the nations that have ratified the protocol. Not all states, however, are ready for prime time.

CHAPTER 2

The Economics of
Climate Change Policy

The intensity of debate on the natural and physical science of global climate change is fully matched on the economic side. Invariably, the Kyoto Protocol and alternative proposals to reduce greenhouse gases have undergone economic scrutiny that produces enormous variation in estimating implementation costs. Economic interpretations of Kyoto, applied to the United States or internationally, range from cataclysmic disruption to smooth adjustment. Until governments actually try to reduce greenhouse gases, of course, there will be little real indication of the accuracy of these projections. But why would state governors and legislators take unilateral policy steps that might impose economic hardship on their citizens—and thereby imperil their own political prospects? In particular, why would they do so when there is little demonstrable public demand for such efforts in most states, and any environmental benefits that might be achieved could be realized decades into the future?

Recent state experience suggests that state officials are hardly indifferent to the economic consequences of their actions. Instead, states that have fashioned greenhouse gas reduction strategies have tended to do so in large part because they believe these actions to be in their long-term and short-term economic interest. State politicians are highly sensitive to concerns about economic development, reflected in massive state government efforts to promote investment within their boundaries. They lack the monetary policy tools of their federal government counterparts, and many are constitutionally prevented from running fiscal deficits. Nonetheless, they have more than compensated with a robust set of economic development initiatives that cuts across partisan and regional divides.[1] In fact, many states have borrowed from the European model of extensive intervention in their respective economies, with initiatives that include extensive loan programs to existing firms or venture capitalists, land and worker-training guarantees in exchange for new investment, and a blizzard of tax incentives and abatements to support diverse forms of development. A number of states routinely offer incentive packages that exceed $50,000 for each new job to prospective developers and engage in active bidding wars with other states to maximize their prospects of winning those competitions.[2]

As in the European case, one can quibble over the effectiveness of these state programs.[3] But until recent budgetary woes brought on by recession, states have generally won high marks in the past few decades for their capacity to balance budgets, retain high credit ratings that have sustained low borrowing costs, and concentrate economic development efforts in sectors that successfully retain existing jobs and stimulate new ones. Virtually every state has developed specialized agencies to promote economic development, with many going so far as to open state offices in foreign capitals to foster long-term economic cooperation. In fact, governors and high elected officials of virtually all states now regularly pursue international travel in search of increased trade.

In recent decades, a number of officials with state executive experience have used their stewardship of their respective state economies as central components of successful campaigns for the presidency. The American governorship has increasingly proved to be a viable launch pad for the White House, as four of the past five presidents—Jimmy Carter, Ronald Reagan, Bill Clinton, and George W. Bush—had sought elected office only at the state level before running successfully for the presidency. All emphasized the ways in which they had promoted economic prosperity in their respective states as a model for their approach to national governance.[4]

In short, state governments—and their elected leaders—are intensively focused on economic development. They are hardly inclined to initiate policies that will weaken their competitive economic edge with neighboring states and other regions of the United States. Consequently, the recent flurry of state activity to address global climate change is especially noteworthy because supportive political leaders have endorsed these steps in part—or in full—because they see them as contributing to long-term economic well-being. But not all states view the possibilities for greenhouse gas reduction solely through an economic lens. Some have indeed come to perceive climate change as a serious environmental threat, one that warrants a significant policy response that seeks to minimize any economic downside of implementation. These are the prime-time cases (see figure 1-1). In a range of state cases, however, economic considerations have been paramount in guiding discussion of climate change, either allowing some experimentation to go forward—however it is labeled—or precluding any activity whatsoever.

These cases reflect the significant differences in issue framing for climate change, which presents markedly differing opportunities for potential policy entrepreneurs depending on the state in which they reside. This variation was perhaps best illustrated in August 1998, when the Environmental Council of the States (ECOS) convened a two-day meeting in Madison, Wisconsin, to discuss possible state strategies to address climate change. The council was formed in 1993 to represent the environmental commissioners of the American states and territories and the District of Columbia. It regularly attempts to share ideas and experiences and to promote interstate collaboration on a range of environmental issues. The 1998 ECOS meeting on climate change was chaired by environmental commissioners from six states, reflecting regional diversity and a mixture of commissioners appointed by both Republican and Democratic governors. "We had long since concluded that climate change was real and that taking steps to reduce [greenhouse gas] releases was hardly a devastating economic issue," recalls one of the state cochairs.

> In fact, we had learned that there was a lot of low-hanging fruit and that we could pick it in ways that were good economically for the state. But I was just stunned when we put the meeting agenda out to the other states. I got some angry calls, including commissioners from some states asking, "Why the hell is this on the agenda? We can't touch this issue." It was then that I realized the constraints officials from some other states are working under on this issue.[5]

Climate Change Mitigation as an Economic Threat: The Hostile States

One of the fiercest critics of that ECOS agenda and most strident opponents of any actions that might reduce greenhouse gases has been the State of Michigan and its Department of Environmental Quality (DEQ).

Michigan ranks eighth among states in total greenhouse gas emissions (see table 1-1), and elected officials at the state level as well as those who represent the state in Congress have long opposed any regulations that might impose more rigorous standards for clean air or fuel economy on the motor vehicle industry. But the state has also had a historic reputation as a national leader on environmental and conservation issues, reflected in a record from earlier decades that featured pioneering legislation in water pollution, recycling, public participation in environmental decisionmaking, and energy conservation.

That record changed markedly during the three terms of Governor John Engler (1991–2003), who pursued a vigorous agenda to promote economic development by attempting to minimize the reach of state government on a range of regulatory and fiscal matters. Engler repeatedly singled out environmental regulation as a detriment to economic well-being throughout his governorship, with any proposed intervention to affect climate change only a further illustration of governmental intervention run amok. During the presidential campaigns in 1992, 1996, and 2000, Engler routinely carried a copy of *Earth in the Balance* and quoted extensively from its climate change section while on the stump for Republican candidates. He decried Gore as an environmental—and global warming—fanatic who "thinks he's smarter than the auto industry, the oil industry, the men and women who build the cars."[6] Such remarks, and related policies to curtail any state initiative to reduce greenhouse gases, were well received by the state's dominant industry, the Big Three auto manufacturers.

Engler dominated relations with the state legislature and presided over a series of major changes in state environmental policy. These included an executive order to abolish nineteen state boards and commissions that oversaw various areas of environmental policy, including air pollution and water resources, and legislation to remove key elements of so-called polluter-pay legislation that had received recognition as a national model at the time of its enactment.[7] Michigan under Engler also moved aggressively into the energy field, reflected in 1999 legislation to restructure the electricity industry. Whereas other states, such as Texas and New Jersey, used this type of legislation to include provisions that promoted energy conservation and renewable energy sources, Michigan moved in the opposite direction. A series of long-standing "demand-side management" programs that the state used to require utilities to promote energy efficiency and conservation were completely eliminated and were not replaced with new initiatives. These programs were labeled discriminatory toward citizens who were ineligible to receive funding from them and inconsistent with the goals of a competitive electricity marketplace that the state was attempting to create through its restructuring legislation.[8] During this period, the state also repealed its energy efficiency codes for residential and commercial buildings, although these codes had long since been standardized in most states and during recent years have been made increasingly rigorous in many.

Michigan also conducted a major reorganization of its environmental programs through a 1995 executive order that divided environmental oversight between the established Department of Natural Resources and the newly created Department of Environmental Quality. This coincided with efforts by Engler to secure more direct oversight over these departments and their directors and to reduce overall state environmental staff through a series of budget cuts. In addition, the state periodically offered generous retirement buyout packages to senior state employees and then replaced multiple retirees with at most one individual, thereby steadily shrinking overall state capacity in environmental management. Many state agency officials who were most likely to evolve into roles as climate change policy entrepreneurs, given prior work in related areas such as emissions trading and energy efficiency, are simply no longer employed by state government.

This was hardly a fertile political environment for agency-based policy entrepreneurs to promote new ideas to reduce greenhouse gases. Clearly, the most effective state environmental policy entrepreneur on climate change—albeit a type of entrepreneurship to block rather than foster new policy—was Michigan's DEQ director Russell Harding. Harding headed the DEQ from its inception in 1995 until early 2003 and was in clear sync with Engler. He developed a national reputation for leading opposition to any state intervention on climate change as well as most other policy initiatives proposed by the federal government or by the states and provinces of the Great Lakes Basin. Under Harding's regime, the DEQ regularly spurned federal grants to study climate change and became an outspoken opponent of any early actions. At the 1998 ECOS meeting on climate change, Harding went out of his way to express concerns about state engagement on this issue and to make clear that any existing state policies that might have some impact on reducing greenhouse gases were in no way intended to achieve that outcome. "We have tried and tried to contact them and get them involved in some way, and so has Region 5 of EPA," notes a member of the EPA staff to support state and local initiatives. "We never get a response. In fact, when we call the department and ask to speak to someone knowledgeable about climate change, the standard answer is that the department simply does not deal with that issue." Harding's opposition focused on the uncertainty of climate science, contending that "the jury is still out on any man-made causes." But his primary emphasis was the potential economic risks of state engagement on climate change. Such risk "is a big issue for Michigan," Harding states. "We don't want to get into a position of telling companies they can't come here."[9]

Michigan took formal steps to make sure that no policy entrepreneurs might still be lurking in state agencies with plans to analyze greenhouse gases or propose ways to reduce them. In November 1997, just before the Kyoto negotiations, the state passed a resolution that decried any future efforts to reduce greenhouse gases as an enormous economic threat. It urged President Clinton to refrain from international negotiations on climate change that exempted developing nations or "that adversely affect the people, prosperity, or employment stability of the United States or any region or sector."[10] In 1999 the state attempted to amend the 1994 Natural Resources and Environmental Protection Act to prohibit any state agency official from taking any steps to "propose or promulgate a rule intended in whole or part to reduce emissions of greenhouse gases, unless the legislature has enacted specific enabling legislation for such a rule."[11] This proposed legislation also included provisions to constrain expenditures from state funds and restrict interactions with federal agency officials that might serve to reduce greenhouse gases. The bill passed the Michigan House and had strong support from the Engler administration but ultimately failed to secure a majority in the state Senate. Nonetheless, its development served to further chill any intrastate consideration of policy to reduce greenhouse gases. "Basically, if you work for an environmental agency in Michigan," explains a senior environmental agency official from a neighboring state that tried—and failed—to work cooperatively with Michigan on a series of environmental concerns, "you are not allowed to do or say anything that could be relevant to climate change."

It is not clear, however, that this aggressive Michigan posture against greenhouse gas reduction will endure in the post-Engler era. The governor consistently won high marks from state voters in public opinion polls for his performance in economic development and earned national notoriety for his role in welfare reform. But his ratings on environmental protection were far lower throughout his term in office, and his environmental record began to be seen as a political liability late in his tenure and in the 2002 election. Even some Michigan-based members of the auto industry began to shift policy gears on this issue, most notably the Ford Motor Company, as did a number of other large manufacturers based in the state. Michigan began to be singled out as being behind the curve on a host of environmental protection issues as the Engler administration neared the end of its final term.

In response, Engler closed his three terms as governor with an about-face of sorts on greenhouse gases. In April 2002 he announced plans to create a major state center for energy research, entitled Next Energy. Despite the fiscal shortfall of more than $1 billion the state faced, Engler proposed an initial commitment of $50 million to launch the center. One of its primary goals, he noted, would be the development of technology that "could render the internal combustion engine obsolete."[12] In his announcement, Engler spoke favorably of long-term alternatives such as hydrogen fuel cells and hybrid engines and said he wanted to make sure Michigan industries were on the cutting edge in developing these new technologies. Although he attempted to differentiate his approach from the "kind of ruinous policy Gore was advocating," he acknowledged that this initiative could produce both environmental and economic benefits for Michigan.

A number of environmental protection and energy issues relevant to climate change surfaced in Michigan during 2002, some of which figured in that year's election. Proposals to create a mandatory program for renewable energy and to integrate carbon dioxide reductions with other conventional pollutants in air pollution legislation emerged for the first time in the state legislature, although both failed to reach Engler's desk where, in any case, a veto was expected. In the 2002 gubernatorial campaign, Engler's handpicked successor, Lieutenant Governor Richard Posthumus, visibly distanced himself from Engler on several key environmental and energy issues. Ultimately, Posthumus was defeated by Democratic attorney general Jennifer Granholm, who emphasized a renewed commitment to environmental protection as one of her major campaign issues. The early months of her administration were dominated by massive budget cuts to balance the state budget, but increased attention was directed to a series of areas relevant to greenhouse gas reduction.

Eleventh-hour conversions notwithstanding, Michigan's actions have reflected an interpretation of the potentially dire economic effects of unilateral action to address climate change. They may reflect the strong views of a dominant industry, although, in this case, they appear more indicative of a powerful governor's more traditional linkage between economic development and the softening of environmental protection. Michigan may well reflect an extreme example of states that fall into the "hostile" camp (cell 5 in figure 1-1). But it is not alone in taking actions to deter policy entrepreneurs from pursuing greenhouse gas mitigation opportunities.

Colorado, for example, also fits into the category of "hostile" states on the basis of a 1998 resolution of the legislature. This provision expressed deep concern over the potentially adverse impacts of greenhouse gas mitigation and responded with an attempt to block any state policies that might attempt reductions. The state has followed a path quite different from Michigan's, however. In fact, considerable potential for incremental policy innovation to reduce greenhouse gases earlier in the decade may have been quashed by overly ambitious and adversarial entrepreneurs—namely, a body of consultants who antagonized a pair of strong Colorado industries, coal mining and coal-burning utilities.

Before this reversal in the late 1990s, Colorado had actively sought and secured more than $100,000 of EPA funding in 1994 to begin an inventory of its greenhouse gas emissions and consider potential policy options through an action plan. The Air Pollution Control Division of the Colorado Department of Public Health and Environment (DPHE) was particularly interested in exploring these issues. The department thought that such analysis might contribute to development of an integrated, long-term strategy for air pollution that could be linked with an active set of earlier state programs to foster energy efficiency and conservation.

But matters changed dramatically once the drafting of an action plan began in 1996 and 1997. Lacking the in-house expertise to conduct this project, the state contracted with Recom Applied Solutions, a private consulting firm based in Longmont, Colorado. A draft report by Recom staff was released in September 1997 and triggered a strong and negative reaction from many quarters of the DPHE and various stakeholders. The report endorsed a bold set of actions, most of which were concentrated on electrical utilities based in the state. More than 98 percent of Colorado-generated electricity comes from the burning of coal, and the state also hosts a substantial coal-mining industry. The Recom proposals called for an across-the-board change in state policy toward utilities. These included plans to phase out older power plants, impose a moratorium on new coal-based plants, and aggressively pursue new energy efficiency programs that would meet 40 percent of the state's new electricity needs over the next two decades. The report also encouraged the state to formally link any restructuring decisions, such as possible state or consumer coverage of "stranded costs," to formal utility commitments to reduce their greenhouse gas releases.[13]

These recommendations were not well received by either the DPHE or the state's utilities. "The coal industry went bonkers," recalls a DPHE official. "They said we weren't presenting a balanced picture and were trying to put them out of work. They went right to the executive director of the department and also threatened to go to the legislature." The DPHE responded by asking the authors to consider modification of the report to include more politically and economically feasible recommendations. But after several meetings, according to a DPHE official, "it was apparent that they were not interested in trying to balance out the information contained in the document or in changing the tone of the document to address our political sensitivities." Instead, the state used some of the data from the consultant report but produced an alternative report that simply offered a menu of possible policy options, most of which were far less likely to be controversial.[14]

The earlier draft report had served to generate such intense controversy over the very idea of state mitigation of greenhouse gases that it ultimately weakened the prospects for any meaningful steps that might be considered. This experience thereby served to shift the state from a more opportunistic stance toward a hostile one. As in Michigan, the legislature in Colorado sprang into action in early 1998 with a resolution that received resounding support in both chambers. The resolution emphasized an economic analysis conducted by Wharton Econometric Forecasting Associates that contended that any effort to reduce Colorado's greenhouse gases below 1990 levels would cost the state more than 29,500 jobs "while subjecting Colorado's citizens to higher energy, housing, medical, and food costs that would reduce Colorado's tax revenue by $420 million." The resolution then implored President Clinton not to sign the Kyoto Protocol and urged the Senate to reject it in the event that Clinton did sign. It also included language that discouraged state or federal agencies from taking "any action to initiate strategies to reduce greenhouse gases as requested by the Kyoto Protocol."[15] This restriction was not as exact or as binding as its counterparts in some other states, both because of its more flexible language and the fact that it only expressed legislature intent as a resolution. In fact, Governor Roy Romer was widely expected to veto the proposal if it reached his desk in the form of legislation.

Nonetheless, the resolution clearly chilled earlier plans in some circles of the DPHE to begin to move ahead on some modest greenhouse gas reduction strategies. As in Michigan, the Colorado case illustrates the capacity of elected officials to frame any effort to reduce greenhouse gases as an economic threat that prompted them to work to thwart any potential entrepreneurial activity.

This resolution has not completely prohibited state engagement on initiatives related to greenhouse gases, whether or not they carry such an explicit label. Unlike Michigan, the state has continued to work with EPA, accepting federal funds to conduct a study on ways the Colorado ski industry might promote pollution prevention, reduce conventional air pollutants, and reduce greenhouse gas releases. This project has been explicit about its goals to measure and attempt to reduce greenhouse gases, in conjunction with a larger effort to maximize the economic efficiency of the industry. The state also has four cities—Aspen, Boulder, Denver, and Fort Collins—that participate in the International Council for Local Environmental Initiatives and have begun to consider their own greenhouse gas reduction programs.[16] New interest in generating and purchasing renewable energy, most notably from wind, is also increasingly evident in several areas of the state through the efforts of private and nonprofit groups.[17] Moreover, Colorado has retained its extensive set of earlier energy efficiency and conservation programs, even expanding these in a few instances.[18] In these cases, there tends to be no explicit labeling of greenhouse gas reduction initiatives, though these programs clearly can reduce overall emissions. "It goes forward, but in a very low-key way," explains a state official. "For us, linking it with other issues is the only way to go." Observers concur that this incremental process may have begun to soften the initial coal-centered hostility toward serious deliberation on this issue that was triggered by the Recom report, facilitating a shift toward discussion of greenhouse gas reduction as an economic development opportunity.

Climate Change Mitigation as an Economic Threat: The Indifferent States

Whereas some states, such as Michigan and Colorado, have taken formal steps to demonstrate their hostility toward proposals that would reduce greenhouse gases, other states have simply made no policy response to the issue. Among the states examined in this study, Louisiana and Florida clearly emerge as "indifferent," reflected in their placement in cell 6 in figure 1-1. Neither state has passed any formal restriction or prohibition on action by respective agencies, and both accepted federal funds to complete an inventory of their greenhouse gas emissions early in the current decade. But neither Louisiana nor Florida has developed subsequent action plans nor enacted any of the sorts of policies evident in prime-time, opportunistic, or stealth states.

In some respects, this indifference is ironic. With the possible exception of Hawaii, Louisiana and Florida have potentially more to lose in coming decades from climate change than any other state. Both have extensive ocean coasts and wetlands, with substantial areas at or below sea level. A series of studies concur that these states are highly vulnerable to a future rise in sea level stemming from global warming. These threats may be especially severe in Louisiana. According to a 2001 analysis, "The deltaic coast of Louisiana exhibits the greatest vulnerability to sea-level rise of the entire U.S. coast." The study concludes that "New Orleans is particularly threatened by the rapid loss of the surrounding wetlands in the Mississippi Delta, which currently protect the city from storm surges."[19] Large coastal expanses of southern Florida present comparable potential threats, and the state has been characterized by one official with a southern Florida planning council as a "canary in the mine shaft of sea-level rise."

Nonetheless, it remains difficult to detect much of a pulse for serious consideration of greenhouse gas issues in either state. This does not represent a continuation of the pattern in states such as Michigan, Colorado, and West Virginia, where officials were attentive to the concerns of carbon-intensive industries and took steps to preclude greenhouse gas reduction initiatives. Instead, both states appear to be overwhelmed by other environmental issues, such as toxics contamination in Louisiana and water protection in Florida, and they have limited experience as leading-edge states on environmental and energy policy innovation. Their recent histories indicate some modest engagement on greenhouse gases, with a possible rise in involvement in recent years.

In 1997 the Louisiana legislature commissioned a report on the possible ramifications of climate change for the state. The report notes that Louisiana could be the "poster state" for climate change, including potentially massive loss of wetlands in the state that contains approximately 40 percent of the coastal wetlands in the nation. Little serious consideration of the issue—however labeled—has occurred since the report was released in 1998. The state has continued to pursue a decade-long strategy of wetland restoration and freshwater and sediment diversion projects, although these are not expected to have much impact if sea levels continue to rise.[20] However, legislation enacted in 2002 will result in the formation of a state commission to examine possible state policies that could mitigate the effects of climate change in the state.

Florida has been equally quiet on this issue, although an anti-Kyoto bill was introduced in 1998. It passed the House Environmental Protection Committee in March 1998 but was never considered on the floor of either chamber. Interviews in both states suggest little interest in this issue among state agency officials or elected officials; most environmental groups have remained riveted on other matters. Perhaps the greatest expression of recent concern about climate change is emerging neither from governmental nor environmental groups but rather from industries that are clearly at risk if sea-level projections are realized. In particular, representatives of the tourism and insurance industries in Florida have encouraged discussion on the issue, but without any policy results to date. This may have contributed to introduction of legislation into the Florida Senate in 2002 that would cap carbon dioxide emissions from the state's electric power plants at 1990 levels by 2007. The legislation was not approved, but its very introduction constitutes a significant shift in the tone of state policy deliberation on this issue.

Climate Change Mitigation as a Stealth Source of Economic Development: Texas Messes with Renewable Energy

The state of Texas would, in many respects, appear likely to follow the examples of states either hostile or indifferent to developing policies to reduce greenhouse gases. The state is well known for its prodigious consumption of energy, consistently ranking well ahead of all other states in terms of total energy and electricity use, including industrial energy consumption and total petroleum consumption. It also continues to be a major power in energy generation, leading all other states in electricity generation and petroleum production. That energy is primarily provided by fossil fuels such as oil, coal, and natural gas, much of it retrieved from beneath Texas soil. Texas ranked dead last among the fifty states in the percentage of electricity it generated from renewable sources in 1995.[21] At the same time, the state has emerged as the national leader in terms of total greenhouse gas emissions (see table 1-1) and total toxic emissions to air, land, and water. It also has some of the most pervasive air pollution problems in the United States, particularly in such major metropolitan areas as Houston and Dallas–Fort Worth.

Regardless of partisan control of state government, Texas has not traditionally been considered a champion of environmental causes. Engagement in greenhouse gases would seem a particular stretch, as no state—or nation—might be more disrupted economically by a major emissions mitigation program, given the state's carbon-intensive economic base. This may explain why Texas has never enacted explicit legislation that addresses greenhouse gases and made no commitment to study the issue until late August 2000. At that point, the Bush administration reversed course right before the presidential election, seeking federal support to enable the Texas Natural Resources Conservation Commission to conduct an inventory and announcing plans to study possible greenhouse gas reductions.[22] Nonetheless, Texas members of the U.S. House and Senate have remained outspoken in their opposition to international greenhouse gas agreements, such as the Kyoto Protocol, as well as more incremental mitigation strategies introduced to Congress, on the basis of anticipated economic repercussions for the state.

But Texas has never been bashful about pursuing economic development, particularly in the field of energy. During the 1990s, the state looked seriously at long-term restructuring of its patterns of electricity generation and consumption. It ultimately endorsed a fairly traditional deregulation program in 1999, but it wove into that legislation a far-reaching plan to mandate that a steadily increasing amount of renewable energy be generated and blended into the state's pool of available electricity. This step was not merely symbolic. Indeed, the provision has triggered dramatic growth in state generation of renewable electricity, primarily through rapid expansion of wind power. This has already begun to reduce Texas's enormous volume of greenhouse gases. However, environmental—and greenhouse gas—considerations were not paramount in securing support for this policy. In fact, the authorizing legislation made no explicit reference to either greenhouse gas or carbon dioxide impacts. Instead, the driving force was economic, as the Texas legislature and Governor Bush concluded that it made economic sense to attempt to diversify the state's sources of electricity.

The Texas case exemplifies the possibilities of stealth approaches to climate change policy. The state did not officially go on record with regard to climate change but seized upon a particular policy that would approach greenhouse gas reduction as good for its economic development. In this case, increased generation of renewable electricity offered such possible economic benefits as expanded supply and improved reliability of access to electricity. This was a particular concern to the state because of its relative difficulty of importing and exporting electricity, given its location on the continental electricity grid. State initiatives suggested the possibility of greater energy independence for the state through reduced dependence on sources from other nations and states. Woven into this package were environmental concerns, including reduction of air pollution and its impacts on quality of life. Policy proponents, including agency-based entrepreneurs who cultivated the idea of expanded use of renewables during the 1990s, were aware of the greenhouse gas ramifications. They consciously decided to play them down, however, and instead accentuate other features.

Electricity, Renewables, and the Texas Portfolio Standard

Electricity generation contributes greatly to American greenhouse gas releases. The burning of fossil fuels for electricity is responsible for approximately one-third of the gases generated in the United States—and in Texas—primarily in the form of carbon dioxide. There are no current technologies available to trap the carbon dioxide generated by electric power plants before it is released to the atmosphere, as there are for conventional pollutants. Consequently, many traditional policy tools for controlling air quality, such as mandatory scrubbers to remove certain pollutants before release, are simply not applicable at present. Instead, as the environmental policy analysts Howard Gruenspecht and Paul Portney have noted, "reduction in energy-related carbon dioxide emissions within the U.S. can only be achieved through policies that reduce the use of energy, or shift the energy mix in favor of low- and no-carbon fuels."[23]

For almost a century, state governments have played central roles in the regulatory oversight of electricity, through state utility commissions that make fundamental decisions on pricing, approval of new facilities and technologies, and promotion of conservation.[24] In Texas, such powers reside with the state Public Utility Commission (PUC), which includes a staff of analysts and administrators who report to a board of commissioners appointed by the governor. The 1992 federal Energy Policy Act posed new challenges for these state government bodies, opening the opportunity to consider electricity restructuring and the possibility of new competition that would involve electricity suppliers who had historically been excluded from the marketplace. This legislation did not force states to restructure but set in motion a series of planning reviews that opened a policy area traditionally closed to new ideas and forces.

Numerous states have responded to this opportunity with legislation that attempts to provide industrial and residential consumers with some degree of choice in selecting an electricity provider. These efforts have tended to be exceedingly complex both technically and politically. They must address issues ranging from access to traditional power lines by alternative power generators to compensation from ratepayers to established utility companies for potential losses of market share. California's troubled experiment with restructuring is, of course, most familiar to Americans and has given other states pause in implementing new initiatives of their own in the past few years. Some states, however, have seized upon restructuring in their efforts to increase their reliance on renewable sources, derived from "naturally regenerated" sources such as the sun and wind.

State governments have used a number of mechanisms in recent decades to foster the generation of renewable electricity, most notably, tax credits and grants. More recently, sixteen states, including Texas, have enacted legislation that includes "renewables portfolio standards" (RPS). These require utilities operating within state boundaries to provide a certain amount or percentage of power from renewable sources as part of their total offering of electricity. As table 2-1 suggests, these programs vary markedly from state to state, both in the levels of electricity required to be derived from renewables and in other key provisions. States that set higher levels tend to already have significant sources of renewables, such as extensive hydro-based power in the case of Maine. The RPS concept originated abroad and is currently used in a range of nations, including Australia, the Netherlands, and the United Kingdom. Some countries have formally integrated these programs into their comprehensive strategies to achieve pledged greenhouse gas reduction targets, as has New Jersey.

Texas created its renewables portfolio standards through the Restructuring of Electric Utility Industry Acts, which received wide support from both chambers of the state legislature and was signed into law by Governor Bush on September 1, 1999. The Texas program mandates "cumulative installed renewable capacity" measured in total megawatts (MW), climbing gradually from 880 MW at the time of enactment to 1,280 MW in January 2003, 1,730 MW by January 2006, 2,208 MW by January 2007, and 2,880 MW by January 2009.[25] Every megawatt of electricity

Table 2-1. *Renewables Portfolio Standards for Electricity, by State, 2003*

State	Share to be met by renewable energy (percent)	Target date
Arizona	1.1	2007
California	20.0	2017
Connecticut	13.0	2009
Hawaii[a]	9.0	2010
Illinois[a]	15.0	2020
owa	2.0	1999
Maine	30.0	2000
Massachusetts	11.0	2009
Minnesota[b]	10.5	2015
Nevada	15.0	2013
New Jersey	6.5	2012
New Mexico	5.0	2002
New York	15.0	2020
Pennsylvania	0.2–2.0[c]	2001
Texas	3.0	2009
Wisconsin	2.2	2011

Source: Data from Barry G. Rabe, *Greenhouse and Statehouse: The Evolving State Government Role in Climate Change* (Arlington, Va.: Pew Center on Global Climate Change, 2002), p. 12.

a. Indicates a renewables portfolio "goal" as opposed to "standard"; no penalty for noncompliance.

b. Indicates a renewables portfolio "goal" statewide but requires that its largest utility, Xcel (formerly Northern States Power), attain this level of renewables. The state recently advanced the date for Xcel to meet the standard to 2006.

c. Varies by utility. Figures are for 2001, increasing thereafter.

provided by renewable sources saves 3,863 tons of carbon dioxide that would have been generated from the existing mix of fossil-fuel sources.[26] Texas anticipates that by 2009 at least 3 percent of its electricity will be generated from renewables, if the target is achieved, although the state appears likely to exceed that amount, given the robust growth of its renewable sources since enactment of the standard. Texas also established a Renewable Energy Credits Trading Program that gives utilities considerable flexibility in meeting the standard requirement. Under the program, any electricity provider that cannot meet the RPS may "satisfy the requirements by holding renewable energy credits in lieu of capacity from renewable energy technologies."[27] Every renewable-energy project in Texas that has been certified generates one "renewable-energy credit" for every kilowatt-hour (kWh) of electricity it produces. These credits can be purchased by electricity providers to meet any shortfall in their own generation of renewable energy.

The Political Path to Renewables

Much like other states with more explicit climate change policies, Texas did not receive a substantial amount of media attention for its RPS, and the proposal did not emerge as a leading priority for most of the established environmental groups operating in Austin. Instead, it reflected nearly a decade of idea development and coalition building, nudged along by a series of entrepreneurs and ultimately inserted into a single page toward the end of a sixty-one-page bill. These entrepreneurs took advantage of a number of opportunities both to raise the issue of renewable energy in general terms and to build on growing concerns about the long-term reliability of the electricity supply. With this foundation, they then pushed to include the standard within the larger legislative package.

This was no small task, given both the carbon-intensiveness of the Texas economy and the tradition of the Texas PUC. Historically, Texas has worked aggressively to develop its own energy sources and has taken formal steps to avoid any engagement with other states or the federal government. It avoided the creation of a public utility commission until 1975, long after such bodies were well established in most other states. Without such governmental oversight, the state actively encouraged aggressive energy development alongside a semiformal separation from the remainder of the North American network for electricity distribution. As figure 2-1 indicates, the United States and Canada operate an electricity "grid" with a series of separate "interconnected systems," most of which include large expanses of territory and clusters of states and provinces. Texas is unique in that the system that provides its power, the Electric Reliability Council of Texas, largely adheres to existing state boundaries.

This separation did not reflect any formal policy; rather, it resulted from an informal agreement by unregulated state electricity generators to maintain as much independence from other states—and the federal government—as possible. "It's just a Texas thing," explains Patrick Wood III, who chaired the Texas PUC under Governor Bush and subsequently served as chair of the Federal Energy Regulatory Commission under President Bush. "We want control of our own destiny."[28] Creation of such a system along state contours makes it particularly difficult for Texas to export its electricity or to import electricity from other systems. In fact, the structuring of the continental system is more conducive to movement of electricity—and collaborative greenhouse gas reduction policies—between a number of American states and Canadian provinces than between Texas and any other states.

Figure 2-1. *Power Pools of the North American Electric Power Grid*

ECAR	East Central Area Reliability Coordination Agreement
ERCOT	Electric Reliability Council of Texas
FRCC	Florida Reliability Coordinating Council
MAAC	Mid-Atlantic Area Council
MAIN	Mid-America Interconnected Network
MAPP	Mid-Continent Area Power Pool
NPCC	Northeast Power Coordinating Council
SERC	Southeastern Electric Reliability Council
SPP	Southwest Power Pool
WECC	Western Electricity Coordinating Council

Source: North American Electric Reliability Council (www.nerc.com/regional [September 24, 2003]).

The delayed introduction of electricity oversight also led to a system in which there were minimal pressures to promote energy conservation or efficiency, even in the 1970s, when many states developed active programs that have, in some instances, been expanded to further reduce greenhouse gases. Moreover, the Texas PUC was widely perceived as a classic model of a "captured" regulatory body, one whose commissioners were eager to meet any demands of the utilities that they were supposed to be overseeing. It demonstrated little concern for providing mechanisms for citizen or environmental group input and still less for long-term planning for electricity supply.[29]

But this detachment began to change in the 1990s for a pair of reasons. First, Texas began to realize that its historic isolation as an electricity provider might not necessarily be as conducive to economic development as it had long assumed. In 1992 Texas became for the first time a net importer of energy, and analyses made toward the end of the decade indicate that the state was consuming 12 percent more energy than it produced.[30] Texas continues to generate prodigious amounts of various fossil fuels, but its production of both natural gas and oil peaked in the early 1970s. A highly influential study on Texas energy issues published in 1998 acknowledges that the state continued to lead the nation in natural gas and oil production but notes that "Texas is among the most thoroughly explored mineral resource provinces on earth." If all wells that had already been drilled "were uniformly spaced," the report observes, "there would be a well every half a mile in every direction throughout the entire state."[31] The report also notes the diversification of the Texas economy in recent decades, demonstrating the steady decline in the percentage of the state work force employed by—or tax revenue derived from—the energy industry. Clearly, the state was not facing an imminent "energy crisis," and its electricity turbines were in no immediate danger of lacking fuel to continue to meet the high level of demand for electricity. But the traditional presumption that state sources could meet all state demand with ease, in isolation from the rest of the nation and continent, was shaken by these revelations.

Second, the growing concerns about supply reliability coincided with federal mandates under the 1992 Energy Policy Act for states to begin to develop long-term strategies for energy, particularly electricity. This was intended to help states prepare for an expected opening up of state and national energy markets to new forms of competition. In 1995 the Texas PUC responded by launching an "integrated resource plan." The PUC and its regulated parties approached the idea of deregulation with some trepidation but agreed to a multiyear planning process. It also issued a general call for some form of public involvement in this review process.

Many observers assumed that this review would be a closed process and that most of the focus in any deregulation bill would be to ensure that existing utilities received generous terms. But the plan's call for some level of public engagement created an unexpected opportunity to begin to draw attention to the option of renewable energy. "There really was not much of a sense of how to deal with the public participation part of this," notes a former state agency official who is widely credited as having been a pivotal policy entrepreneur behind the RPS initiative.

One night, one of the commissioners of the PUC watched this program on PBS. There was this professor from [the University of Texas at Austin] talking about "deliberative polling" and how this could be used to get a real sense of what the public really wanted. The commissioner knew that the PUC had to do something with this public participation requirement, and he put the [integrated resource plan] idea together with the idea of deliberative polling. We all just ran with that as fast as we could.

Deliberative polling is the brainchild of James Fishkin, a political theorist at the University of Texas. The PBS program reflected his many years of writing and advocacy on behalf of a more iterative and contemplative approach to measuring public sentiment. Under traditional polling, diverse and statistically significant numbers of citizens are asked questions on the spot; their responses are then tabulated to give a sense of the public's view, informed or otherwise. Under deliberative polling, a large number of diverse citizens are gathered in a common place for multiple days of study and interaction on various issues. These participants are asked to read various background materials and consider presentations from diverse perspectives. They are also encouraged to ask questions of the presenters and to interact with one another. At the end of this multiday process, according to Fishkin and other proponents of deliberative polling, a more informed public view emerges.[32] Participant responses to an identical set of questions asked at the beginning and the end of the sessions are used to gauge whether there has been any shift in opinion once citizens have learned more about a given subject.

Such polling has primarily been used on a trial basis in select election campaigns in the United States and the United Kingdom, but PUC staff arranged for deliberative polling sessions within the territory of every major electricity provider in Texas. Between two hundred and three hundred customers were selected randomly for each session and invited to attend at utility expense. Participants were asked to read a "fact book" on electricity generation and come prepared for a weekend of deliberation over the future of electricity in Texas. "We brought Fishkin in as the opening speaker, and he would speak about Athens and democracy and set the general tone," recalls a former state official who helped organize the sessions. But that was followed by "a presentation from every sector with a dog in the hunt" of Texas electricity generation, according to an industry representative who participated in the sessions.

Conveners drew on a number of sources of information, including research on renewables and the Texas energy situation that had been commissioned by the Texas State Energy Conservation Office and the Texas Sustainable Energy Development Council. Governor Ann Richards had decided against releasing an Energy Development Council report on renewable energy in 1995 until her reelection battle against George W. Bush was over. "Then she lost the election, went off to an island with a bruised ego after the defeat, and the study gathered dust in a warehouse," recalls one Texas official. "But there was data there, and it proved very useful in the next stages. All of this information meant that the participants would not just get the opinions of the utilities."

As these sessions proceeded, somewhat surprising findings began to emerge. Utility company representatives and members of leading citizen groups had assumed that the price of electricity would prove to be the top concern and that the sessions could yield valuable insights. For utilities, this information would guide their strategy on pricing and securing maximum benefits in any future restructuring deal. For citizen groups, it would support their efforts to press for reduced costs for residential customers when restructuring was on the table. Instead, the findings from the deliberative sessions confirmed that reliability and stability of support were considerably more important to participants than price. They clearly favored long-term stability with relatively few price fluctuations. Moreover, participants proved unexpectedly responsive to proposals to develop renewable energy, even if the cost might be slightly greater than that from traditional sources. This reflected strong participant support for minimizing environmental damage through the generation of electricity. However, participants "were not volunteering to pay very much" more money for renewable sources and also concluded that, given technological constraints, any introduction of renewables would have to occur gradually.[33]

Findings from these polls proved influential as the PUC and the legislature prepared to consider restructuring proposals during 1998 and 1999. "The PUC commissioners were involved in a panel at the end of each polling session," a state agency official recalls.

> It was probably the first time they had ever been in front of that many people and certainly the first time they had ever heard this kind of stuff. Before, they dismissed any environmental concerns or ideas about renewables as a special interest. This gave them star power, and they really liked it. Pat Wood was impressed; he saw this as something the public wanted once they knew what was involved.

An informal coalition emerged to try to push something specific on renewables into evolving restructuring legislation. This began with general endorsements of renewables and possible support for a "green pricing" provision that would allow citizens to purchase renewable electricity if it became available. But momentum continued to build for something more binding, leading ultimately to the portfolio standard. Support came from key officials in the PUC and related state agencies, working with organizations that had some previous involvement in efforts to develop renewable energy in the state or were eager to become involved if new opportunities were provided. The principal environmental group involved in the negotiations, Environmental Defense, was represented in key discussions by an individual who had previously worked in Texas state agencies and as a legislative aide. Environmental Defense was seen as a viable partner by the utilities, given its prior involvement in emissions trading, its work in other state restructuring legislation, and its nomination of a trusted individual to represent its position. The proposal for renewables portfolio standards moved ahead alongside a series of emissions reduction provisions for conventional pollutants such as sulfur dioxide and nitrogen oxide, all of which were included near the end of the restructuring bill. Most environmental groups were focused on provisions for these pollutants, putting pressure on the Bush administration to move beyond its initial proposals for a voluntary program and instead approve binding requirements.

All participants in these negotiations ultimately endorsed the bill as a win-win strategy that could foster new competition in electricity, protect existing utilities through financial coverage of their stranded costs, and also make new commitments on improved air quality. Observers concur that Bush was not heavily involved in the detailed negotiations but that he effectively used the legislation to burnish both his economic development and environmental credentials. "If anyone thinks that this happened without George Bush knowing about it, they are incredibly naive," explained an energy industry official who was involved in all phases of the restructuring legislation. "The governor was very active and aware of this." The RPS plan was portrayed as a response to the deliberative polling, characterized as a way to ensure long-term electricity stability and potentially contribute to a cleaner environment. It was clearly not labeled as a strategy to reduce greenhouse gases. "How you interpret the bill depends on who you are," notes a state official who was active in all stages of the negotiations.

If we had characterized this as something to do with greenhouse gases, it would have hurt the bill's chances. So we didn't. The fact that no one used that term to argue for the bill shows it would not have sold. Now, of course, things have changed a bit. Now people look at this bill and they see what Texas has done on climate. Now, it's a big thing. But not then.

The Texas Wind Rush

Although only in its first years of operation, the Texas renewables portfolio standard has been hugely successful. The program calls for an increase of 2,000 MW of new renewable sources by 2009 while retaining the base of 880 renewable MW from various sources that were in operation when the legislation was enacted. By early 2002, the state was already well ahead of its 2003 target, having added 913 MW through nine new wind facilities (see table 2-2). This includes 279 MW of new capacity at the King Mountain Wind Ranch in Upton County, which is now the world's largest wind-power site. This initial expansion would make Texas the world's sixth-largest national or subnational generator of electricity from wind. The state is currently reviewing a significant number of additional proposals for renewable facilities, primarily involving wind power but also including solar and hydro sources as well as methane gas. Much of its wind capacity is being developed in western Texas, which has exceptionally good conditions for wind power, given the ferocity of regional winds and vast expanses of open space.[34] In fact, Texas ranks second only to North Dakota among all states in its physical potential for wind-generated electricity.[35] The state has also experienced few of the objections to wind-tower siting—such as the appearance of the towers and dangers that turbines may pose to some birds—that have complicated their construction in some other parts of the country

The first comprehensive analysis of RPS implementation, published in late 2001 by the Lawrence Berkeley National Laboratory, confirms the development of the Texas "wind rush," noting that wind power is proving to be surprisingly cost-effective in Texas, compared with traditional sources. Its authors conclude that "RPS compliance costs appear negligible, with new wind projects reportedly contracted for under 3 cents per kWh, in part as a result of a 1.7 cent per kWh production tax credit [established by the 1992 federal Energy Policy Act], an outstanding wind resource, and an RPS that is sizable enough to drive project economies of scale."[36]

Table 2-2. *Operational Texas Wind Power, 2002*

Project	County	Size (MW)
Before Senate Bill 7 (September 1999)		
Big Spring I	Howard	34.32
Big Spring II	Howard	6.60
Delaware Mountain	Culberson	30.00
Fort Davis	Jeff Davis	6.60
Southwest Mesa	Upton, Crockett	74.90
Texas Wind Power Project	Culberson	35.00
Total		187.42
After Senate Bill 7		
AEP Clear Sky Wind Park[a]	Pecos	160.50
Hueco Mt. Wind Ranch at El Paso	Hudspeth	1.32
Indian Mesa	Pecos	82.50
King Mountain Wind Ranch (1)	Upton	76.70
King Mountain Wind Ranch (2)	Upton	2.60
King Mountain Wind Ranch (3)	Upton	200.00
Llano Estacado Wind Ranch at White Deer	Carson	80.00
Trent Mesa	Taylor	150.00
Woodward Mountain Wind Ranch	Pecos	159.70
Total		913.32
Overall total		1,100.74

Source: Texas Renewable Energy Industries Association, *Texas Wind Plants* (Austin, Tex., 2002).
a. Includes the 25.50 MW project formerly known as Indian Mesa I.

Texas is on track to exceed the 2009 MW standard by the middle of the decade, a fact that has prompted consideration in Austin of elevating RPS levels for the end of the decade. "Ultimately, Texas will be the nation's largest renewable energy source," notes one industry official with close ties to President Bush. "We have always been an energy state and will always be an energy state." In fact, Texas officials talk about wind energy as a huge target for future economic development for the state. "What we've done still tends to get dismissed because we're Texas, not California or Massachusetts—the Green states," one industry expert explains.

But last year, we brought in more than half the new sources of wind opened in the whole country. We have not, of course, done what the mavens of global climate change want, passing a law that says we will reduce CO_2 by x percent a year and forcing industries to do things. Maybe more can happen if you don't go in with a heavy hand but if you approach things from a market standpoint.

Ironically, Texas officials now contend that their biggest concerns about future development of wind power hinge on federal actions. First, they acknowledge that long-term extension of the federal production tax credit may be crucial to sustaining wind-power momentum in Texas. Second, they note concern that their isolation in the national electricity grid may prevent them from effectively exporting their wind power outside the state. One of the strongest supporters of the Texas RPS, former PUC chair Pat Wood, is now responsible for trying to secure nationwide transmission as the head of the Federal Energy Regulatory Commission and thereby increase the prospects for Texas to be able to export its latest source of energy.

Climate Change Mitigation as a Stealth Source of Economic Development: Georgia Reduces Motor Vehicle Use

Few sectors are as important to climate change as transportation, which is responsible for approximately 26 percent of total U.S. anthropogenic greenhouse gas emissions. Much like electricity generation, motor vehicles burn fossil fuels that release massive amounts of carbon dioxide. No technological fix is currently available to capture or contain those emissions. Instead, increased fuel efficiency or reduced vehicle use are the primary options for achieving significant reduction of emissions in this sector.

At the same time, no sector may be so impervious to policy innovation at the state level, given the strong American affinity for driving cars and trucks and the traditional reliance on the federal government to set regulatory standards for vehicular fuel efficiency. It is particularly difficult to envision certain states that are economically dependent on the manufacture of motor vehicles, such as Michigan, taking any serious steps to reduce vehicle use or promote greater fuel economy. State transportation departments may face a further disincentive to innovation, given their primary focus on maximizing highway construction and repair. "These departments are set up as pavement agencies," notes a Georgia state official who works closely with that state's Department of Transportation. "They have always had a clear mission, and it is hard for them to think about anything else." These factors converge to explain why states have proved more reluctant to launch greenhouse gas reduction initiatives in this sector than in others that involve industry and electrical utilities. Nonetheless, there are some exceptions to this pattern and perhaps the beginning of a shift in state thinking on this issue, reflected in California's

July 2002 legislation to establish carbon dioxide standards for motor vehicles.

An increasingly common state practice has involved programs to reduce auto and truck usage and, in turn, foster greater use of mass transit or telecommuting. Americans collectively traveled approximately 4 trillion miles in cars and trucks in 2000, as opposed to only 46.6 billion miles on mass transit; reductions in vehicle use thus pose a significant opportunity for states to make a dent in total greenhouse gas releases from transportation.[37] Georgia has become extremely active in this area, as reflected in a series of initiatives developed in the past half decade. As with the Texas RPS, these policies follow a stealth pattern, representing a response to economic development needs that makes no explicit reference to their potential for significant reduction of greenhouse gas emissions. "One of our staff once referred to greenhouse gas reductions in a talk [with program participants], and people nearly fell off their chairs," explains one state official active in transportation program implementation. "In this part of the country, we just don't talk about climate change."

In fact, Georgia has no explicit policies on greenhouse gas reduction. Both the state Senate and House considered anti-Kyoto resolutions in 1999 and 2000. One version included a section prohibiting "either a regulatory, statutory, or policy proposal that calls for reductions in greenhouse gas emissions" in a manner consistent with Kyoto from reaching the Senate floor. It did not, however, rule out "voluntary proposals" that would reduce such emissions "without compromising the economic security and well-being" of Georgia and the nation.[38] Neither resolution was approved, but both served to discourage agencies from taking any active— or explicit—steps to reduce greenhouse gases. In the transportation area, notes a state environmental official who is involved with these issues, "if this program was touted as an initiative for climate change, I don't know if it would be supported. It has that potential, but we just don't talk about it."

Although climate change has not emerged as an explicit agenda item in Georgia, transportation policy and related problems have long been top-tier concerns. The state has responded with a series of voluntary programs intended to meet important environmental and economic development goals. The thirteen-county area that constitutes metropolitan Atlanta grew from about 1 million people in 1960 to about 4 million in 2000. It continues to expand at a rate of approximately 150,000 people a year. The region was developed with little coordinated planning for transportation and has become a textbook example of the problems related to sprawl.[39] For many years during the past decade, this region led the nation in number of vehicle miles traveled daily in commuting from home to work, averaging more than 40 billion miles a year. "We have a lot of people who have a lot of cars who drive a lot of miles," notes one state transportation official.

By the mid-1990s, this staggering volume of transportation began to be recognized as a huge public policy problem. The Atlanta area developed a reputation for declining air quality, reflected in its classification by EPA as a "serious non-attainment area" for ground-level ozone and its dismal rating in many national air-quality rankings. This endangered Georgia's supply of federal highway funds, which can be withheld under federal law if certain air-quality levels are not attained. These funds are among the most popular transfers from Washington to state capitals, as they translate directly into new highway construction and jobs; states are extremely sensitive to any potential loss of these dollars. At the same time, elected and corporate leaders in Georgia became concerned about the overall quality of life in the region and its possible long-term economic impact. National and local media began to run articles about sprawl, poor air quality, and declining quality of life in the region. As one state official has explained, "There was concern, from the governor on down, that economic development would be lost" if this pattern continued.

Neither the governor nor the state legislature pushed for formal legislation that would address this problem. Instead, they registered their concern and turned to various state agencies to take the lead in devising a response. Through an interagency agreement, the Georgia Departments of Natural Resources and Transportation developed a multifaceted strategy and gave significant responsibility for implementation to the nonprofit Clean Air Campaign (CAC). This organization was formed in 1996 and involves a coalition of more than seventy groups representing the public and private sectors, as well as environmental, educational, and public health interests. The CAC receives 80 percent of its funding from the Georgia Department of Transportation, through federal Congestion Mitigation and Air Quality grants. These grants are matched by other sources of state funding and corporate contributions. The CAC works with the Georgia Environmental Facilities Authority, an "instrumentality" of state government, to implement many of its initiatives. Under the Georgia constitution, such instrumentalities are attached to a state agency and have a board appointed by the governor. The Environmental Facilities Authority, for example, is attached to the state's Department of Community Affairs.

This relationship provides somewhat greater autonomy than would placement within a traditional department and may therefore be more likely to foster innovative approaches. In this case, the state was eager to find a mechanism outside the Department of Transportation, given the department's reputation for promoting—rather than limiting—motor vehicle use. "We were very careful in setting it up this way, rather than run everything through a single government agency likely to have a narrow focus," notes one senior state official who has worked closely with the CAC and the Georgia Environmental Facilities Authority. "The business community has really bought into this, and we have been able to build a broad base of support." The CAC also works with several transportation management associations that provide related services to smaller sections of the region, such as the area surrounding Hartsfield International Airport.

The CAC emphasizes a range of voluntary initiatives based on a six-point behavioral model (awareness, attitude, participation, satisfaction, utilization, and impacts) designed to raise public awareness on the issue and ultimately reduce reliance on traditional transportation modes. Some of its core themes build directly on the region's experience in hosting the 1996 summer Olympics, for which the region and state experimented with a number of voluntary transportation initiatives for the first time. "There were enormous gridlock concerns as the Olympics approached," recalls a Georgia Environmental Facilities Authority staff member. "So we experimented with all of these new strategies, and they worked well. Air emissions went down dramatically during the Olympics despite all the visitors. It was then that we realized we were really on to something."

Much of the CAC's early focus has involved public information, most notably an aggressive mass-media campaign to link air quality with traffic congestion and begin to encourage consideration of alternatives to traditional practices. Campaign ads feature prominent Georgians, including governors, professional athletes, and corporate leaders. This type of activity is increasingly being supplemented with a range of voluntary programs that emphasize consultation with public agencies and private entities to tailor individualized strategies to reduce vehicle use. These consultative services explore such options as expanded use of mass transit, the development of biking and walking options, carpooling and vanpooling, and development of flexible working hours to keep drivers off the road during high-congestion travel periods. In one instance, CAC outreach efforts led to negotiations between management and laborers of a General Motors automotive assembly plant in Doraville on increasing the use of mass transit. The deliberations led to the realization that the plant's second shift ran too late to allow workers to use transportation options, leading to a company decision to alter closure time to facilitate a major increase in the use of mass transit. "When we meet with employers, we talk about this in terms of transportation and employee satisfaction, and that really sells," explains one official from the Georgia Environmental Facilities Authority. "It helps them save money, and the clean air comes for free."

The Clean Air Campaign also offers a series of regionwide services, including a free ride-matching program that can be accessed by phone or on the web. For employers who participate in ride-matching or other programs designed either to pool van or car use or to increase the use of mass transit, the CAC's Guaranteed Ride Home program offers a free taxi ride anywhere in the region to an employee of a participating firm in the case of an emergency. The CAC also offers an extensive set of instructional materials that can be used in area classrooms and provides live traffic reports online, including extensive use of the Department of Transportation's "navigator traffic cameras."

The state has included a strong evaluation component, both to gauge awareness of the issues and to assess the impact on transportation practices and emissions. Officials at the CAC and the Department of Natural Resources note that total vehicle miles traveled per person in the region has declined by an average of one mile a day between 1998–99 and 2000–01. A 2002 evaluation concludes that these state efforts have helped move Georgia considerably closer to its goal of securing ozone-level attainment status from EPA by the middle of the decade through demonstrated pollutant reduction. For 2001, the state estimates that a reduction of 476,000 vehicle miles traveled a day can be attributed to these state programs.[40] This translates to an estimated reduction of 79,333 tons of carbon dioxide for that year, although the state does not publicize this aspect of the program. "Right now, we are focusing on other aspects of the program, but all the folks who are working on this know the greenhouse gas impacts are there," explains one state official.

> This is a big nut to crack; it's going to take a long time. But we're beginning to see the climate issue emerge in other ways for the state. As our air people keep talking with other states like Tennessee, North Carolina, and South Carolina about multipollutant strategies, they are still focusing mainly on conventional pollutants. But,

off to the side, they're talking about carbon dioxide and how to bring that in.

Carbon Sequestration and Economic Development Opportunity: Carbon as a Nebraskan Cash Crop

Not all states pursuing policies that reduce greenhouse gases but do so principally for other reasons rely on the stealth strategies employed in Texas and Georgia. In some instances, a state may explicitly label its intent to mitigate the release of carbon dioxide and related gases while taking such action primarily on the basis of perceived economic development opportunities. Such states may not be indifferent to potential environmental benefits from these efforts, but averting climate change is secondary, at best, to anticipated economic advantages. These "opportunistic" states may prove eager to establish infrastructure to prepare themselves for maximum participation—and future economic benefits—from marketing any credits stemming from greenhouse gas mitigation.

If Texas is finding that it has an abundance of wind to market, states such as Nebraska are discovering that they may have considerable potential to store substantial quantities of carbon in their vast agricultural lands. Ironically, Nebraska began to explore the possibilities of enacting "carbon sequestration" legislation and taking a lead role nationally in this area at the very time that the one of its U.S. senators, Chuck Hagel, was the state's most visible spokesperson on the issue. Hagel emerged as a leading critic of the Kyoto Protocol and has remained a prominent opponent of most proposals for early action to reduce greenhouse gases. In the late 1990s, Hagel may well have reflected general sentiment in his state. "There was so much anger in the state over Kyoto that it slowed things down and made it harder for us to move ahead with our own legislation," notes a former state agency official. "A lot of it reflected skepticism about global warming, a lot of it was antiglobal, and a lot of it was anti-Clinton." But the views of Senator Hagel and any public doubts about the veracity of climate change science did not deter Nebraska in 2000 from enacting legislation that explicitly linked agricultural policy with greenhouse gas reductions.

Nebraska's policy makes clear its interest in better understanding the impact of its various agricultural practices on atmospheric levels of carbon dioxide and preparing for future initiatives that might reduce those levels.

Its primary interest in pursuing this policy, however, is clearly economic development. "When you get into global warming, you get into controversy after controversy," explains an official of the Nebraska Department of Natural Resources who has been heavily involved in this issue. "But for us, the questions have been pretty straightforward: Is there a buck to be made for Nebraska? Will there be markets for our carbon? Can we get ready to play in those markets?" Interviews with diverse stakeholders confirm these emphases. "Once you get into the agencies, there is support that warming is real and that this kind of policy could help," notes one prominent farmer active in the sequestration policy formation process. "But among farmers and ranchers, few have put much weight on the warming issue. The real purpose of this is to bring some additional income to Nebraska farmers, whether or not they believe in global warming."

In general, industrial activity has tended to overshadow natural resources in many debates over future climate change initiatives. This is particularly true in developed countries such as the United States, where less than 3 percent of the economy is based on agricultural activity. Nonetheless, American agricultural practices contribute 40 percent of total U.S. releases of methane, primarily from rice and cattle production, and 68 percent of nitrous oxide derived from fertilizer use.[41] These emissions have substantially greater potential for global warming per ton than identical amounts of carbon dioxide.[42] However, their overall impact is dwarfed by the vastly greater volumes of carbon dioxide released into the atmosphere from multiple sectors.

The Nebraska initiative thus far is focused exclusively on carbon dioxide, examining ways in which state farmers and ranchers might play a significant role in mitigating climate change through increased sequestration of carbon in farmland. "Additional amounts of carbon can be sequestered in soils by relatively minor changes in agricultural practices," note the analysts Richard Adams, Brian Hurd, and John Reilly. "'Growing carbon' on agricultural lands would create a new crop for farmers."[43] According to U.S. Department of Agriculture studies that have been widely utilized in Nebraska, American cropland has the potential to sequester more greenhouse gases each year than the combined annual emissions totals of Michigan, Louisiana, and Georgia (see table 1-1).[44]

Agricultural states such as Nebraska are alert to these possibilities. They increasingly view carbon sequestration as a dual strategy to promote better soil conservation practices and facilitate future participation in national or international carbon markets. Substantial amounts of carbon are naturally stored in agricultural soil, but much of this can be released by tillage and other conventional farming practices. Numerous techniques exist for increasing carbon-storing capacity, including "conservation tillage," which seeds and controls for weeds without using a plow. Instead, planting involves creating a slit in soil, placing a seed into the slit, and then restoring the ground to its original condition. Such tillage has been a popular reform for many years because it reduces the total operating costs and soil erosion, with enhanced carbon sequestration another factor supporting its expansion.[45] There are also other ways in which farming practice can be altered to increase carbon storage capacity, including various approaches to harvesting, crop selection, and crop rotation as well as conversion of marginal farmland to forests.

State-sponsored analyses confirm that Nebraska stored substantially more carbon before sodbusting began in the early nineteenth century and that tilling practices during the middle decades of the past century markedly reduced this capacity. But they note that conservation reforms of recent decades, including reduced emphasis on aggressive tilling and efforts to reduce soil erosion, have increased sequestration capacity. A March 2002 study conducted by the Natural Resource Conservation Service concludes that "the application of sound conservation practices in Nebraska cropland is sequestering carbon and is equivalent to an offset of 12 percent of Nebraska's 1999 fossil fuel carbon emissions."[46] This and related studies concur that additional modification of state farming practices could significantly increase this offset potential.

Nebraska became the first state to formally acknowledge this potential and attempt to link agricultural policy with greenhouse gas reduction. Legislative Bill 957 was approved by the unicameral legislature with only one dissenting vote and signed into law by Republican governor Mike Johannes on April 10, 2000.[47] The legislation noted the importance of preparing Nebraska to become engaged on this issue and established the Carbon Sequestration Advisory Committee (CSAC) with a diverse membership that represented various sectors of agriculture, energy, and state government. It also authorized a transfer of funds from the Nebraska Environmental Trust Fund, which distributes proceeds from the state's lottery. These funds were subsequently matched by other sources, such as the Nebraska Farm Policy Task Force, the Nebraska Corn Board, and the Nebraska Public Power District, to fund initial CSAC efforts.

The CSAC continued to build on earlier efforts on the part of the Farm Policy Task Force to study the issue and commissioned a pair of Department of Natural Resources reports. The first of these, released in December 2001, provides a detailed assessment of the scientific and policy issues surrounding future development of a sequestration program. This includes discussion of a series of agricultural reform practices that could increase sequestration capacity and also provide other benefits such as decreased use of fossil fuels, protection of long-term soil productivity, improved water quality, and reduction in off-site sediment damage.[48] The second report, released in March 2002, provides a detailed baseline survey for cropland and grassland on a county-by-county basis.

These reports converge to offer a two-part process to guide the next steps in policy development for Nebraska. They establish the framework for pilot projects that would generate carbon credits and possible trading within one or more of the state's twenty-three Natural Resource Districts. These districts cut across Nebraska's ninety-three counties, organized around river basins rather than political boundaries. They have long provided a main venue for state and local officials to work with the agricultural community on soil, water, energy, and related concerns. Three districts have already expressed interest in serving as pilots and would work with the baseline data provided in the 2002 report.

The CSAC could also be the basis for a permanent Nebraska entity that would "play a leadership and organizational role in carbon related issues."[49] The December report endorsed either sustaining the CSAC as currently constituted or developing a task force on climate change and greenhouse gases. In either case, the proposal recognizes the need to maintain an institutional "lead unit" for the state on agricultural carbon sequestration and related issues, to build upon the strong initial interest expressed by the agricultural community, and to take advantage of future policy development opportunities. These and other recommendations were under review by the Nebraska Legislature in 2003.

Perhaps the biggest challenge facing the Nebraska program is funding, as there are currently no additional sources of state funds allocated either to institutionalize the CSAC or to implement pilot projects. Nebraska, like many other states, is facing an enormous fiscal shortfall, and sequestration proponents are actively exploring funding sources, including possible federal grants through the 2002 federal Farm Security and Rural Investment Act. Ironically, the lack of fiscal resources seems a more serious impediment to further state action on climate change in this case than any shortage of political support. The appeal of this approach to greenhouse gas reduction is further demonstrated by its rapid diffusion to other states.

Although they did not consult with Nebraska officials, Illinois, North Dakota, Oklahoma, and Wyoming passed strikingly similar versions of the sequestration legislation during 2001.[50]

Nebraska, of course, did not have another state from which to borrow a carbon sequestration program. As in other states, the policy formation process was fairly quiet and received virtually no attention from state media; there were no mass demonstrations by Nebraska farmers or environmental groups clamoring for state support for carbon sequestration. Instead, Nebraska's homegrown approach emerged gradually, reflecting the collaboration of policy entrepreneurs employed either by the state or in agriculture.

The very idea for such a policy, in fact, may well have stemmed from a letter sent in 1998 from the office of Senator Bob Kerrey to the Nebraska Department of Agriculture. "One day we got this letter from Senator Kerrey, exploring whether we had any possible interest in making a link between agriculture and sequestration," recalls Cyd Janssen, who served at the time as an assistant secretary of the state's Department of Agriculture under governor (now U.S. Senator) Ben Nelson. "He was looking for ways to possibly amend the new farm bill to pay farmers for sequestering carbon."[51] Janssen was a member of the Nebraska Farm Policy Task Force, which was formed, and supported financially, by a variety of state farm organizations to examine ideas to generate new markets—and income—for Nebraska farmers. The task force had seventeen members and was cochaired by prominent leaders of the state's farming and ranching communities.

Initially, Janssen was advised against presenting the sequestration issue to the task force, given the roiling controversy over climate change and the Kyoto Protocol. "But I took the issue to them, they listened, and we moved ahead," she recalls. As a result, Janssen was sent to a conference on climate change held at Ohio State University in July 1999. One of the presentations, by John Brenner of Colorado State University, used Nebraska as an example of the potential sequestration capacity of agricultural reform. "I began to see this as something that could improve air, water, people's health, *and* could be a source of money," recalls Janssen. "Usually, the only way farmers can make money is to harvest a resource. This was a way to provide payment to conserve those resources."[52] Moreover, many of the proposed farming reforms to increase sequestration reflected practices that Nebraska had already been implementing or was trying to expand to address other concerns, such as soil erosion.

Janssen's conference report triggered extensive discussion of the sequestration issue within the task force. "It all came back to her attending that
conference," notes one of the task force members. "I didn't know much
about this stuff until then." Initially, the task force struggled with the
issue and whether to take any steps. "There was some initial enthusiasm,
but some of the farmers thought that this was a liberal, communist thing,"
explains one state official active with the task force and in the development
of the policy. "Some were clearly afraid that if they bought into carbon
sequestration, they were buying into the theory of global warming, which
would mean more regulations and forcing them to change a lot of things.
But as we looked at it we could see how it might work."

An important turning point was a conference on sequestration hosted
by the task force at the University of Nebraska. This included the same
presentation from Professor Brenner that Janssen had heard at Ohio State,
as well as presentations from state government experts and other aca-
demics, including policy analysts from the University of Nebraska. More
than a hundred people attended the conference, which gave new credibil-
ity to the idea of pursuing carbon sequestration. "We got a lot of smiles
and head-shaking, with folks asking: 'What the hell are you trying to do
with this global warming stuff?'" recalls one of the task force cochairs.
"But after the University of Nebraska meeting, they began to understand,
and things began to move ahead."

The task force provided a strong supportive coalition for developing a
legislative proposal. Senior officials from the Nebraska Departments of
Natural Resources and Agriculture, along with the Nebraska state con-
servationist and supportive legislative staff, contributed key components
that went into the bill. Senator Merton "Cap" Dierks proved an invaluable ally and championed the bill through the unicameral legislature.
Dierks, a veterinarian with fifteen years of legislative experience, used his
role as chair of the Senate Agriculture Committee to address various concerns and secure an unusually rapid transition from bill development to
enactment.[53] "The senator participated in a number of the early meetings
and was fairly convinced that the carbon markets would develop," recalls
one state agency official. "He wanted Nebraska to be ahead of the game
and was also concerned that we take steps to avoid any sucker payments."

Indeed, concern over possible fraud in any future carbon credit arrangement provided further impetus to Nebraska's actions. The task force had
heard reports that a Canadian consortium representing various utility
companies had begun to negotiate with individual Iowa farmers, seeking

long-term agreements that would compensate them in exchange for transferring sequestration credits. "There was some fear that groups like this would go door-to-door and start signing up farmers with bogus agreements," one Department of Natural Resources official recalls. "We wanted to be sure to avoid this situation in Nebraska."

This concern contributed to the strong emphasis on analytical rigor, especially the need to secure precise measures of existing land-use practices and their impact on carbon. Supporters felt that this had to be done retroactively, to the extent possible, to enable farmers to make the case for credits for conservation initiatives they had undertaken before 2000. "Farmers were concerned that if they'd done good conservation in the past that they might be penalized," another official from the Department of Natural Resources explains. "Some even threatened to tear up their fields and start all over again, if that's what it took to get credits. We doubted this would actually happen but wanted to make sure it didn't."

Supporters also wanted to ensure that the legislation was forward-looking, creating mechanisms to encourage farmers to take steps to increase carbon sequestration in the future. "Soil conservation is extremely important for Nebraska—and it also happens to offer benefits for global warming and our farmers," explains a Department of Natural Resources official. "So, for us, here's another reason to support something we already believe in. We see it this way, not as some landmark environmental bill." Observers concur that Nebraska's first greenhouse gas initiative could lead to more such efforts, not only in agriculture. One very real possibility is expanded collaboration between agriculture and electrical utilities. In particular, the Nebraska Public Power District, the state's publicly held and dominant supplier of electricity, has been actively exploring greenhouse gas reduction issues and considering possible collaboration with Nebraska farmers. Such a coalition would reflect a further expansion of state policy engagement on greenhouse gas issues and perhaps move Nebraska toward prime-time status. "What is carbon sequestration? A couple of years ago, I couldn't pronounce it, much less tell you what it was," notes a task force member. "Global warming is not what's driving this policy, but our farmers are increasingly accepting the global-warming arguments. We're looking at the warmer winters and the changes in growing conditions. They're starting to take this more seriously." Indeed, some states have been taking this issue very seriously for some time now.

CHAPTER 3

The Mechanics of
Climate Change Policy

Economic considerations come into play for any state that contemplates a policy response to the challenge of climate change. No state wants to enact policy that harms existing economic activity or deters future investment. But unlike opportunistic and stealth states, a significant number of states have framed greenhouse gas reduction policies largely as a response to the perceived environmental threat posed by climate change. These states tend to have studied the issue for some time and considered a range of policy options. They have responded with multiple policies that are explicit about their role in attempting to reduce greenhouse gas releases. In many respects, these prime-time states, by developing policies and supportive administrative infrastructure, have become serious players in climate change policy. They understand the ways in which they generate greenhouse gases and have begun to tailor policy responses accordingly.

Prime-time states began to go on record a decade or more ago in acknowledging concern over climate change and in beginning a process of study and policy experimentation. They have consistently framed climate change as a significant environmental challenge for their state, noting current indicators of climate change and highlighting potential impacts. In response, prime-time states tend to have established units—albeit small in some instances—in lead agencies that provide continuing leadership on the issue and guide implementation of new policies once they are enacted.

Four of the twelve states in this study clearly fit the prime-time category. New Hampshire, New Jersey, Oregon, and Wisconsin have each launched multiple policy initiatives related to climate change, reflecting an extended period of engagement on this issue and considerable nurturing by policy entrepreneurs. Many elements of Illinois's response to climate change also fit the prime-time category, although it is a more complex case that also demonstrates features of other categories. At least one-fourth of the thirty-eight states that are not included in this study appear to warrant inclusion in the prime-time cell, as well. These include a diverse group, ranging from the remaining New England states and New York to Minnesota in the Midwest and California and Washington on the West Coast, among others. It does appear that there is a general trend toward prime-time approaches among states, consistent with the significant growth in the number of state climate change policies. In fact, some of the opportunistic and stealth states appear to be shifting toward prime-time status, just as some hostile or indifferent states may be considering taking initial steps toward problem analysis and policy development.

The prime-time states examined in this study represent a curious mixture. They tend to have very differing economic bases; New Jersey and Illinois have long been known for their concentrations of heavy industry and substantial generation of conventional pollutants, whereas New Hampshire and Oregon have featured mixed economies with smaller-scale industrial development. Wisconsin, with its combination of industry, agriculture, and tourism, falls somewhere in between. These states also reflect very different patterns of partisan control, having been governed by a mixture of Republican and Democratic governors and legislatures during the past decade. Indeed, no one political party appears to have a monopoly on climate change policy leadership at the state level. In turn, some of these states—such as New Jersey, Oregon, and Wisconsin—have traditionally been national leaders in environmental policy innovation, whereas others—such as New Hampshire and Illinois—have been less likely in recent decades to assume such roles.

Despite their differences, these states have tended to frame climate change in similar ways, as an environmental threat that warrants a serious policy response. They have been willing to label many if not all of the policies that they institute as expressly dedicated to reduction of greenhouse gases, most commonly carbon dioxide. Collectively, they offer a sampling of the range of policies available to polities—subnational or national—to combat climate change.

No two states have responded with an identical set of policies. But all of the prime-time states have tempered their efforts to reduce greenhouse gases with a strong recognition of economic consequences. Like their opportunistic and stealth counterparts, these states have attempted to fashion policies that make long-term economic sense while putting themselves on a path to mitigate greenhouse gases and prepare for even greater policy involvement in the future. In many instances, they have won the support of the industries and utilities that will be most directly affected by accentuating the advantages that stem from early action. These include offers to adjust other regulatory provisions in exchange for greenhouse gas reductions, opportunities to claim credit for reductions, and clarity on state regulations across various categories of pollutants that will allow firms to plan for the future. As New Hampshire governor Jeanne Shaheen, who signed that state's carbon dioxide legislation, has noted, an increasing number of states "are discovering solutions that provide their communities with extraordinary collateral benefits at the same time, including saving money, improving quality of life, and enhancing economic development and competitiveness."[1] Just as states are getting ready to participate in the long-term policy arena of greenhouse gas reduction, many regulated firms have been willing to join in this process for similar reasons.

The emergence of some states as prime-time players has been evident for several years, perhaps initially reflected in the debate that took place among the state environmental commissioners in the 1998 ECOS meeting in Madison, Wisconsin. A number of states were so reluctant to engage this issue that they expressed opposition to the very idea of discussing climate change in this context. Nevertheless, the meeting went forward, and a number of states signaled their commitment to move ahead actively, developing new strategies tailored to local economic realities. As Robert Shinn, the director of the New Jersey Department of Environmental Protection, noted at the ECOS gathering,

> I am here today to tell you that taking action to reduce emission levels does not have to conflict with our overall economic growth. In fact, I am convinced that an early effort to improve energy efficiency and bring to market renewable-energy technologies will enhance our economic prosperity. Delaying action will only add to the cost of reducing emissions while increasing the instability of the global climate in the next century and beyond.[2]

Such sentiments have been expressed by a growing number of states, including the ones introduced in the balance of this chapter.

Linking Greenhouse Gas Mitigation with Air Pollution Regulation: New Hampshire

States can influence greenhouse gas releases from electricity generation not only through their oversight of utilities but also through their roles in setting and enforcing air pollution standards. Federal laws, most notably the 1990 Clean Air Act Amendments, impose numerous requirements that states are required to implement. Nonetheless, states have continued to play an innovative role, with clusters of them long committed to taking action to combat air pollution from either industrial or mobile sources that is sometimes subsequently embraced as federal policy. Just as the "California effect" has long guided auto emission standards, and may do so once again with that state's 2002 legislation on carbon dioxide emissions from vehicles, other states took the lead in finding mechanisms to combat sulfur dioxide emissions well in advance of the 1990 legislation.[3] Wisconsin and New York were particularly active on this issue during the 1980s, helping pave the way for the 1990 law as well as preparing themselves for subsequent engagement in greenhouse gas mitigation.

Consequently, it is not surprising that some states are taking the lead in defining the next generation of clean-air policy by expressly incorporating carbon dioxide into multipollutant strategies. In contrast to the Bush administration, a growing number of states have begun to define multipollutants to expressly include the largest source of greenhouse gases, in addition to more conventional pollutants such as nitrogen oxide, mercury, and sulfur dioxide. Massachusetts, for example, took formal action for operational power plants in April 2001, when Republican governor Jane Swift issued a rule for a multipollutant cap that includes carbon dioxide for six major power plants. "The new, tough standards will help ensure older power plants in Massachusetts do not contribute to regional air pollution, acid rain and global warming," said Swift in announcing her support for the regulations.[4]

New Hampshire followed Massachusetts in May 2002 with legislation that also applies a multipollutant cap to carbon dioxide and other pollutants. This reflects a decade of in-state policy development, including earlier initiatives that set the stage for the legislation. The New Hampshire Clean Power Act received bipartisan support from the state legislature and was signed into law by Governor Shaheen, a Democrat. The legislation requires that the state's three existing fossil-fuel power plants stabilize their carbon dioxide emissions at 1990 levels, which is approximately 3 percent below their 1999 levels, by December 31, 2006. Two of these plants, located in Bow and Portsmouth, use coal, and the third, located in Newington, uses oil and gas. All three plants are owned by Public Service of New Hampshire, a division of Northeast Utilities that also operates plants covered by the Massachusetts carbon dioxide rule.[5] To achieve the mandated reductions, the plants must either reduce their generation, increase fuel efficiency, or purchase emission credits from other plants outside New Hampshire that have achieved such reductions, or use some combination of these strategies.

The Political Path toward an Air Pollution Regulation Program

New Hampshire might appear to be a particularly unlikely candidate to develop such an active and explicit program for greenhouse gas reductions. It ranks forty-fourth among the states in terms of total greenhouse gases (see table 1-1), meaning that any reductions it achieved would represent only a small portion of national totals. Its small size could indeed raise questions about the economic desirability of taking such bold unilateral steps were it to develop a reputation for imposing regulatory burdens not evident elsewhere. Moreover, the very idea of New Hampshire leadership on climate change regulation seems a contradiction in terms, given the state's historic emphasis on minimizing governmental interference through any form of regulation or taxation, reflected in the state's famous motto, "Live Free or Die." As the environmental journalist Tom Arrandale has noted, "Among all the states, New Hampshire hasn't usually ranked as a trailblazer on protecting the environment."[6]

New Hampshire has, however, been sensitive to the impact of air pollution, particularly sulfur dioxide emissions leading to acid rain. The state has felt the environmental impacts of, as is commonly stated, "being on the end of the tailpipe of America": climatic patterns often lead conventional pollutants from the south and west to emerge as New England air-quality problems. State-based industries and agency officials, in turn, have felt that New Hampshire did not receive fair treatment under the 1990 Clean Air Act Amendments, with insufficient credit being given for early state reductions of sulfur dioxide emissions.[7]

A pair of well-placed policy entrepreneurs built on that experience to launch a decade-long campaign for early state action on climate change. Both Kenneth Colburn and Robert Varney held senior positions in the New Hampshire Department of Environmental Services (DES) during the

1990s and formed a solid team that enabled the department to take the lead on a series of climate change initiatives, culminating in the 2002 legislation. They cultivated potential allies, including legislators, governors, and representatives of industry and environmental groups, in securing support for new policies.

Kenneth Colburn directed the Air Resources Division of the DES, remaining in that post until shortly after the New Hampshire Clean Power legislation was enacted. He was highly active in a number of national and regional associations and developed a national reputation as an innovative leader on climate change and air pollution issues. Colburn was clearly thinking about new approaches that could influence climate change well before they became politically feasible in New Hampshire. One such proposal, the Industry-Average Performance System for air pollution, involved a bold initiative for expanded use of market mechanisms in developing a new approach to air-quality protection. It was written with state representative Jeffrey MacGillivray and received a 1997 award for better government ideas from the Josiah Bartlett Center for Public Policy.[8] The Industry-Average proposal never worked its way into legislation, but many other ideas that Colburn cultivated did. "He was all over the place, representing New Hampshire in places like Rio and the Hague," one former colleague recalls. "He definitely knew CO_2; it was a hot-button item for him."

Robert Varney served as DES director during this period and "was just passionate about climate change," according to one industry representative. He provided Colburn with political support and resources to begin a multiyear and multistep process of framing climate change as a serious concern to New Hampshire and advancing the case that early reduction efforts would make both environmental and economic sense. "Colburn and Varney were an incredible team," notes one former associate. "They started with this informal educational process on climate change, talking to those elements in the state interested in learning more about the subject. You could just see things starting to turn."

Preparing the Case for Action

New Hampshire began its efforts by actively pursuing federal grants and developing analytical capacity to measure greenhouse gas emissions and consider policy options. The state was not reluctant to label its efforts as focused on climate change. It recruited climate change specialists to its Department of Environmental Services, often using federal grants to cover

expenses. The DES developed a series of sophisticated publications outlining recent changes in New Hampshire climate patterns and possible
threats that climate change posed to key industries. "Before, New Hampshire had a cold shoulder on climate change, and some legislators
expressed skepticism about us doing anything," explains one state official.
"But what helped us in convincing them was stressing the economic
impacts of climate change on New Hampshire. Loss of the maple sugar
industry to Canada because our maple trees were migrating north got
some legislators involved; dangers to tourism, especially the ski industry,
from reduced snowfalls got some ski-district legislators going." Concerns
about tourism, from altered recreational opportunities to possible loss of
colorful leaf-turning seasons, were particularly significant, tourism being
the second-largest industry in the state. The DES proved particularly effective at developing estimates of the economic impacts on the state of those
sectors particularly vulnerable to future climate change.[9]

Colburn became known nationally for his advocacy of a coordinated
approach to reductions in air pollution, including that caused by carbon
dioxide emissions. He was an active player behind an influential 1998
study published by the State and Territorial Air Pollution Program Administrators and the Association of Local Air Pollution Control Officials,
which singled out three metropolitan areas and one state—New Hampshire—for analysis. The study concludes that state and local governments
could indeed fashion "a menu of harmonized options" capable of achieving integrated reduction of greenhouse gases and conventional pollutants
and that "technically feasible and cost-effective" strategies were indeed
"well within the reach of most states and localities."[10]

This applied analysis reinforced general research findings on the potential "corollary benefits" of addressing multiple pollutants simultaneously.
"We regulate pollutants one by one," Colburn said in a 2002 address.
"But the impacts are integrated."[11] It set the stage for future action,
although Colburn, Varney, and allies had to wait for the right opportunities to move ahead. The first possible legislative opportunity, involving
electricity restructuring, occurred in May 1996, but this was much too
early for the state to take meaningful action on climate change. However,
the DES did succeed in raising long-term concerns about greenhouse gases
and helped secure support for a social-benefit charge. This provision
involved a charge of two mills ($0.002) for each kilowatt-hour for every
electricity ratepayer in the state. Proceeds were to be divided between
state energy efficiency and low-income assistance programs. Thirteen other

states, including Illinois and Wisconsin, have enacted similar provisions. These are known as "mini carbon taxes" in some states owing to their ability to raise electricity costs (and presumably deter fossil-fuel consumption) at the same time that they allocate revenues to promote reduced energy use. In 2001 the state increased the charge to three mills a kilowatt-hour, with all of the additional revenue devoted to energy efficiency programs.

The impact of this program has been limited thus far, however, by protracted litigation over the terms of electricity restructuring. Much of this controversy has concerned stranded costs and provisions for dealing with the controversial Seabrook nuclear power plant and was not formally resolved until 2001. This also delayed further greenhouse gas policy initiatives that involved state-based utilities, although entrepreneurs did advance plans to establish a statewide registry to credit early reductions of greenhouse gases. This approach built upon the federal registry established in the U.S. Department of Energy during the first Bush administration to create a two-tiered system to "help New Hampshire entities take credit for mitigation actions they have already taken."[12] Under this system, participants can claim reductions associated with particular projects, such as converting a boiler from coal to natural gas, or a comprehensive review of all facilities and products associated with a firm. Despite the lingering controversy over restructuring, the statewide registry was approved with broad support in the legislature and signed into law in July 1999. Colburn and Varney played key roles in the development of the registry idea and emphasized the state's commitment to "stand beside sources that had voluntarily made early [greenhouse gas] reductions in good faith."[13] Rules for the registry were issued in February 2001, and the DES has launched an outreach effort to encourage participation and assist companies with technical assistance in ensuring registration of reductions.

Carbon Dioxide under a Multipollutant Umbrella

Like the social-benefit charge, the registry was a prelude to the larger goal of a multipollutant regulatory strategy that included carbon dioxide. During her 2000 reelection campaign, Governor Shaheen was asked by major environmental groups to pledge to take steps to address pollutants from grandfathered plants if elected to a second term. "The governor pledged to do something if reelected, then after election asked the Department [of Environmental Services] to develop a strategy," a DES official explains. "We . . . came forward with the multipollutant strategy with all four pollutants—including carbon dioxide." The previous years of analytical work conducted by Colburn and DES staff made development of the bill relatively straightforward. They emphasized the economic advantages to regulated firms of "being able to look at all pollutants at once" and began to meet with various constituencies to make the case for the bill. Utilities were concerned about some of the provisions, including the levels of proposed mercury and carbon dioxide reductions included in the first version. Some environmental groups opposed provisions for emissions trading, reflecting philosophical objections to any purchase or trade of credits for carbon dioxide or other emissions. Additional questions were raised about the timetable for phasing in required reductions, particularly given the anticipated selling of power plants once the restructuring litigation had been settled.

Ultimately, the DES helped orchestrate a compromise on all major points, leading to overwhelming support. The carbon dioxide reductions were trimmed to achieve stabilization at Rio-like 1990 levels, giving power plants a bit more latitude in compliance, although the date for compliance was moved from 2007 to 2006, shortening slightly the time frame before which a "phase two" of carbon dioxide reductions could be considered. The idea of a second round of reductions was linked to the state's formal involvement in a collaborative plan with other New England states and eastern Canadian provinces to achieve extensive greenhouse gas reductions by 2012. Emissions trading provisions were retained, but greater incentives were provided to minimize reductions within or near state boundaries. These arrangements were sufficient to garner the support of plant operators and reflected their active engagement in the entire process.[14] "This meant they would have some regulatory certainty about the future," a state official explains. "This way, now that the litigation was settled, they could sell their plants and consider long-term options like the role natural gas would play in the future." For some environmental groups, doubts over emissions trading remained, although two major organizations in the state, the Audubon Society and the Society for the Protection of New Hampshire Forests, provided a strong endorsement. Elected officials from both political parties sought to claim credit for the legislation. "This bill sets a national model for how environmentalists, power producers and lawmakers can work together to address serious concerns like air pollution," stated Representative Jeb Bradley, a Wolfeboro Republican who helped shepherd the legislation through the four-hundred-member New Hampshire House and used his involvement in the bill in his successful 2002 campaign for U.S. Congress.[15] Governor Shaheen noted growing evidence of climate change in New Hampshire in announcing her decision to sign the legislation. She explained that climate change "threatens skiing, foliage, maple sugaring, and trout fishing—all crucial to our state's economy."[16] Shaheen used her role in this legislation and the registry to burnish her environmental credentials in her ultimately unsuccessful 2002 campaign for the U.S. Senate. Interestingly, the most outspoken critic of the legislation was Shaheen's potential rival, Senator Robert Smith, who was defeated in his bid for Republican Party renomination by Representative John Sununu, who then defeated Shaheen in the general election.

Public utilities remain in the early stages of determining how they will comply with the new provisions. New natural gas plants are expected to make inroads on the older plants, and many observers anticipate that the Portsmouth-based coal-burning Schiller facility, which has provided electricity to New Hampshire residents for nearly half a century, will be closed. According to the DES, the average annual emission rate of carbon dioxide per kilowatt-hour from the state's existing fossil-fuel-burning power plants is approximately three times greater than the average annual emission rate for new, combined-cycle power plants that burn natural gas.[17] This makes the option of switching fuels particularly attractive, given the advanced age of the existing New Hampshire plants.

Officials at the DES are already looking ahead to a second phase of carbon dioxide reductions and are considering further reduction steps. These include increased emphasis on renewable energy, including wind and solar power, forestry programs to increase carbon sequestration, and expanded state emphases on energy efficiency. However, it remains uncertain whether past momentum will be sustained, Colburn and Varney having left New Hampshire. Colburn "hung on long enough to see the legislation through" but then departed to become the director of the Northeast States for Coordinated Air Use Management, based in Boston. This organization is an interstate association for the air-quality control programs in the eight northeastern states. Varney has become the administrator of EPA's New England regional office, also based in Boston. Although now removed from their New Hampshire bases, from these new venues Colburn and Varney may provide further entrepreneurial opportunities for regional action.

Linking Greenhouse Gas Reduction with the Siting of Energy Plant Facilities: Oregon's Carbon Dioxide Standard

States not only can control greenhouse gas emissions from established power plants (as New Hampshire has done) or the level of renewable energy required for their total electricity pool (as has Texas). Under their extensive powers of utility regulation, they can also establish the core rules that guide the process of deciding whether electricity-generating facilities can be opened or expanded, and they can determine the conditions under which any new siting might be authorized. These conditions may include formal requirements to restrict the levels of greenhouse gases that can be generated by any new facility. Oregon used these powers in the late 1990s to simultaneously streamline its outmoded energy-facility-siting process and formally link any future siting initiatives with explicit commitments to reduce greenhouse gases.

Such a linkage reflects both the expanded state use of existing authority to mitigate greenhouse gases and the rapidly changing world of electricity generation. The opening of new electricity-generating plants was traditionally part of a closed state regulatory process, often dominated by facility proponents who had established intimate working relationships with the state public utility commission. The creation of a new plant or expansion of an existing one was frequently a pro forma exercise. Indeed, local communities that served as "hosts" for proposed plants tended to frame them as attractive economic development opportunities. This was true even of new nuclear power plants opened during the 1950s, 1960s, and through much of the 1970s. In such cases, local communities—and their state regulatory commissions—welcomed new facility siting for its anticipated economic benefits.[18]

One of the few early constraints imposed by many states on new siting proposals was some demonstration by proponents that new electricity generation was necessary to meet anticipated in-state demand. This reflected historic state perception of electricity as an intrastate good and a desire to protect existing suppliers from unwanted competition. Many states responded with "need-for-facility" standards to make certain that their overall capacity to generate electricity closely matched expected state need. Oregon enacted a law in 1971 establishing such a standard, as did many other states across the nation. This policy was similar to other regulatory programs of the period, such as "certificate-of-need" requirements for hospital construction or expansion, in which state governments sought to attain something approaching equilibrium between supply and demand by controlling the development of new capacity.

By the 1990s, such regulation in energy policy began to appear outdated. States contemplated deregulation and began to view electricity on a more interstate and regional basis. Given the option available to most of them for exporting electricity to neighboring states or regions, states might consider development of capacity that exceeded expected in-state need. Because of the uncertainties attending deregulation and energy supply on a national basis, states might also find it attractive to increase their supply to guard against any future disruption of existing sources. Oregon, for example, relied on hydropower generated by dams for more than 80 percent of its electricity during the 1980s and 1990s. This gave the state a significant source of renewable energy, enabling it to reach its relatively low per capita greenhouse gas emissions in 1999 (see table 1-1). But as many Oregon dams began to approach the end of their operational lives and political pressures to return rivers to their natural states increased, the long-term viability of hydropower became increasingly suspect.[19] Oregon also experienced enormous controversy around the operation and ultimate closure of its 1,110 MW nuclear power plant, the Trojan facility, in Rainier. Despite numerous operational problems and cost overruns, a series of state ballot propositions in the 1980s and the early 1990s failed to close the facility. However, it was ultimately shuttered in 1993, after the state approved generous terms for recovery of stranded costs for the plant's owner. Consequently, Oregon, like many other states, became highly interested in diversifying its electricity sources and proved receptive to the possibility of opening new facilities that might not meet a traditional—and immediate—"need-for-facility" standard.

States were not prepared, however, to quickly abandon these well-established provisions. Many environmental groups were staunch supporters of any regulations that could block new electricity-generating capacity, fearful that repeal would simply expand supply and lead to more environmental degradation. Existing electricity suppliers in a state did not necessarily welcome the prospect of added competition in the form of new power plants located within state boundaries, given their historic domination of those markets. In particular, these new competitors might generate less-costly electricity and produce less environmental contamination in the process through use of new technologies and energy sources, challenging the continued existence of established sources. The very idea of energy-facility siting had long since moved beyond its earlier acceptance as an economic development strategy, often facing many of the forms of local opposition inherent in the NIMBY ("not in my back yard") syndrome so common for controversial facilities.

These factors converged in the 1990s as Oregon, like many other states, struggled with the issue of injecting competition into the electricity sector, revamping its traditional regulatory process while advancing environmental protection goals. This was not an easy transition, and ongoing controversies continued in many states. But a series of policy entrepreneurs in Oregon understood the competing concerns and steadily advanced an idea to link a more flexible regulatory posture with formalized commitments to expanded environmental protection in the form of greenhouse gas reduction. This entailed a series of provisions that ultimately led to 1997 legislation, which eliminated the long-standing need-for-facility standard. In exchange, a new standard requires that any new or expanded power plant proposed for operation in Oregon attain a level of carbon dioxide emissions that is 17 percent below the most efficient natural-gas-fired plant currently in operation in the United States. Proposed facilities may meet this standard in one of two ways: through development of new technologies that achieve greater efficiencies than existing practice or through purchase of carbon dioxide offsets by making monetary contributions that underwrite carbon mitigation projects.

The Political Path to a Carbon Dioxide Standard

Oregon's interest in developing new energy sources coincided with a strong concern about climate change as a serious environmental problem. The state has long had a strong commitment to environmental protection, and it was the first state in the nation to enact a variety of policies, ranging from a "bottle bill" that placed a deposit on all bottles and cans sold in Oregon to statewide antisprawl legislation. A constant theme in its transition from a natural-resource-based economy to a high-technology "new economy" has been emphasis on environmental quality to sustain Oregon as an attractive place to reside and invest.[20] The state has long been highly active in promoting energy conservation and efficiency, giving considerable resources and authority to the Oregon Office of Energy. For example, the office has operated a self-supporting loan program since 1981 to assist both private and public entities to either increase energy efficiency or use renewable energy. Loan capital funds are provided by state bonds, which may be issued to a level of $40 million annually. Those loans made to private institutions can then be matched with Oregon's 35 percent tax credit for business energy efficiency.[21] Numerous state energy efficiency incentives are focused at the level of individual consumers. For instance, Oregon provides a tax credit for purchase of appliances that exceed federal efficiency standards by at least 25 percent.

This historic involvement on environmental and energy issues paved the way for early engagement by the state on climate change policy, and it is one of the reasons Oregon has earned categorization as a prime-time state. Oregon officials began to frame climate change as a serious environmental concern for the state in the late 1980s, when the Office of Energy chaired a task force involving twelve state agencies to study the potential impact of climate change in Oregon and possible state responses. This initial review set the stage for inclusion of climate change in a major process of strategic planning for the state's future. The planning process, called Oregon Shines, placed a strong emphasis on the state's long-term health and prosperity, establishing five, ten, fifteen, and twenty-year goals in three categories: people, economy, and quality of life. Oregon Office of Energy staff made a successful case for the inclusion of specific carbon dioxide reduction goals as an important part of this process, beginning with a Rio-like goal of holding state emissions at 1990 levels. Progress in achieving these goals continues to be measured and made public through a related program of key state performance measures, known as Oregon Benchmarks. This program has attained considerable state and national visibility and has been a supportive force behind new policies designed to help achieve the state reduction goals.

The inclusion of carbon dioxide at this early stage also enabled the Oregon Office of Energy to expand its initial involvement on climate change. The office published a long-term state energy plan in 1993 that outlined possible paths toward attaining Oregon Benchmark goals on carbon dioxide reduction and pursued federal grant support that underwrote a detailed inventory of state greenhouse gas emissions and policy options. The state quickly realized that population and economic growth during the 1990s would make it impossible to reach its Benchmark goals, thereby contributing to expanded exploration of policy tools that might reverse the trend. One early step, which would contribute to the ultimate adoption of the carbon dioxide standard, was a 1993 decision by the Oregon Public Utility Commission to require state-based utilities to estimate the externality costs of their emissions. This was added to earlier PUC provisions requiring more sophisticated utility analysis of estimated costs for existing and proposed plants. The commission specifically included "externality adders" for three conventional pollutants and carbon dioxide. As a result, "any generation proposed had to include their emission as a quantified cost borne by the public."[22]

Regulatory Trade-offs

By this time, "work on climate change had become somewhat conventional" in Oregon, according to Sam Sadler, a senior energy analyst from the Oregon Office of Energy who was instrumental in all the steps leading to the 1997 legislation.[23] This created an opportunity to attempt the integration of greenhouse gas reductions with likely changes in state utility regulations. By the mid-1990s, utility industry representatives were clamoring for repeal of the need-for-facility standard. But environmental groups opposed abandonment of the standard, and Governor John Kitzhaber expressed his opposition to outright repeal. Instead, officials from the Oregon Office of Energy worked closely with the state's Energy Facility Siting Council, which oversaw siting proposals, to advance plans for a one-time exemption from the standard.

This compromise gained the support of the state legislature and was signed into law by Kitzhaber in 1995. It enabled the state Energy Facility Siting Council to conduct a competition that would allow the siting of one facility with a capacity of up to 500 MW. "The idea of a one-time competition for the right to build a new plant appealed to a lot of different groups," recalls a former state official who now works in the private sector. The successful applicant would receive its approval based on its capacity to demonstrate the lowest level of environmental impacts. The primary criteria for measuring these impacts would be minimization of the proposed facility's "net monetized air emissions per kilowatt-hour, including carbon dioxide."[24]

After an extensive review process, the exemption was granted to the 305 MW Klamath Cogeneration Project, to be located in Klamath Falls in southern Oregon. The proposal featured a state-of-the-art combustion turbine and a $4.9 million commitment to purchase "carbon dioxide offsets" through solar rural electrification, methane use at sewage treatment plants and coal mines, reforestation of underproductive agricultural lands in western Oregon, and expansion of the geothermal heating district for the city of Klamath Falls. These offset commitments reduced the total carbon dioxide emissions from the new facility, which, because of the new technology, were already well below the national averages, by 28 percent. The combined offsets over a century and related reductions by the plant during the life of its expected operation are expected to achieve a carbon dioxide reduction of 10.3 million metric tons (MMT), equivalent to more than 90 percent of the state's total greenhouse gas emissions for 1999 (see table 1-1).

The success of this one-time experiment set in motion a search for a long-term strategy to link energy development with carbon dioxide emission reductions. The governor and legislature created a seven-member Energy Facility Siting Task Force, with staff support from the Oregon Office of Energy, to study possible next steps. The task force, which included a number of members either currently serving in the legislature or with past experience in that body, came forward with a unanimous recommendation for a quid pro quo whereby the need-for-facility standard would be eliminated in exchange for carbon dioxide standards that would be applicable to any new or expanded natural-gas-based power plant in the state. The standard effectively eliminated the possibility of siting plants that burned fossil fuels with higher carbon dioxide emissions, such as coal, because of its application to a cleaner fuel, natural gas. "Over time, support had built to do something," recalls a state official who was heavily involved in the process. "So we set up this trade-off between need-for-power and CO_2 standards. Environmental groups said yes to the CO_2 standards; industry said yes to the end of need-for-power."

This proposal received relatively little attention in the state media; it passed both chambers of the Oregon legislature with limited discussion and without an opposing vote and was signed into law by Kitzhaber in June 1997. "I am not sure the legislature really understood what they were voting on, as it was just too complicated for them," an industry representative explains. "But these state people had been at it a long time, more than a decade in some cases. They had worked to build support, and it showed." Another observer, a member of a major environmental group involved in energy issues in Oregon and neighboring states, has noted that "there are a couple of key officials in Oregon who have been instrumental in making a lot of this happen. They have been at it quite a while, know their stuff, have lots of interaction with other agencies, utilities, local governments, and environmental groups. They are connected and respected, and that helped make this thing go." Yet another observer, a senior state official involved throughout the process, recalls, "We were prepared. When it became possible to address climate change in this area, we were ready."

The creation of a carbon dioxide standard 17 percent below best performance of any facility operating in the nation led to the initial requirement that the proposed plant "shall not exceed 0.70 pounds of carbon dioxide emissions per kilowatt-hour of net electric output." It also allowed state officials "to reset the carbon dioxide emissions standards" if they discovered that new technologies were in operation anywhere in the nation that achieved greater reduction of carbon dioxide emissions.[25] Consequently, the construction of a more efficient plant in Vancouver, Washington, enabled the state to revise the standard to 0.675 pounds of carbon dioxide emissions per kWh in January 2000, retaining the statutory objective of remaining 17 percent below best national practice.[26]

Pursuing Offsets

Attainment of such a steep reduction level necessitated creation of a mechanism to secure credits for carbon offsets through other projects, as had occurred in the original Klamath case. The legislation recognized this need through creation of the Oregon Climate Trust, which was authorized to "purchase offsets with funds provided by power plant developers."[27] The trust is a nonprofit organization governed by a seven-member board with representation from state government, industry, and environmental groups. As in the Georgia transportation case, the process for project review and fund distribution was intended to foster public trust and remove this activity from the control of an existing state agency. The trust is required to spend at least 80 percent of its funds on direct carbon reduction projects, with the remainder used to cover administrative costs as well as project monitoring and evaluation. Projects are approved through an open competition that can include proposals from other states and even other nations. Decisions are based on a variety of factors, but most important are "whether the Trust funding will make a significant difference in whether the project goes forward or not, and the project's realistic potential for CO_2 reductions."[28]

The first project approved under the new legislation involves an expansion of the Klamath Cogeneration Project through the construction of an additional gas turbine that will add nearly 200 MW to the facility's output. In addition to using the most advanced technology available, the facility has agreed to provide cogeneration steam to an area lumber company, to halt use of oil as a backup source to natural gas except in cases of emergency, and to contribute $1.2 million to the Oregon Climate Trust for carbon mitigation projects. The trust subsequently completed a competitive review process and announced a number of project grants in 2001 and 2002. These included purchase of old growth and other forests for removal from the industrial forestry cycle, creation of web-based ride-matching programs, and support for new wind and landfill gas energy projects. These projects are collectively intended to reduce or capture at least 764,300 MMT of carbon dioxide over periods that range from ten to one hundred years for the various projects.

The trust was designed to be able to pursue projects outside the state of Oregon, and in 1999 its name was formally changed to the Climate Trust. The trust has established a cooperative project with Seattle City Light, the city's municipally owned utility, to pursue carbon mitigation projects related to purchase of electricity from the Klamath facility.[29] It authorized its first international project in May 2002, a contract with Ecuador's Jatun Sacha Foundation and Conservation International to capture and store atmospheric carbon in a rainforest. This will entail planting native hardwood trees on a 680-acre site in the northwest coastal mountains of Ecuador, a rainforest rated by Conservation International as one of the world's five most threatened areas of biodiversity. The project is intended to sequester at least 65,000 MMT of carbon dioxide over a hundred-year period.

Just as the Climate Trust is cutting across jurisdictional boundaries, the state's carbon dioxide legislation has already had spillover impacts on other sectors. This has probably been most notable in the growing linkage between regulation of electricity and promotion of forestry. Separate state legislation in 1997 allowed state-based utilities to include the costs of tree-planting programs into their rates, providing for cost recovery to give utilities "experience in offsetting CO_2 emissions" through forest carbon sequestration.[30] The Oregon Department of Forestry has a long-standing interest in the potential carbon sequestration effects of expanded forestry planting and protection programs. The legislature established the Oregon Forest Resource Trust in 1993 with an initial authorization of $3.5 million to develop more-robust forestlands and explore possible links with carbon storage. This level of funding has not been sustained, but the Forest Resource Trust has begun to work cooperatively with the Climate Trust on carbon mitigation projects, including a joint contribution of $1.5 million for carbon offsets through the Klamath project.

The state moved beyond this initial program, however, through July 2001 legislation that set up a mechanism "to market, register, transfer or sell forestry carbon offsets on behalf of the landowners to provide a stewardship incentive for nonfederal forestlands."[31] This legislation authorizes the Oregon State Forester, in the Department of Forestry, to develop a forestry carbon-offset accounting system for the state and holders of forestlands. Much like the Nebraska effort to promote agricultural carbon sequestration, the state is clearly attempting to establish a structure to support initial measurement of sequestration potential and facilitate initial projects and trading. One very real possibility would be a multijurisdictional partnership, such as developing comparable legislation in nearby states and Canadian provinces that establishes a regional framework for developing such an area.

Oregon has also continued to explore greenhouse gas reduction policies in other areas, such as transportation, consistent with its prime-time status. The state's extensive set of tax credits and related programs to encourage businesses to invest in transportation efficiency projects, for example, have helped make Oregon one of the heaviest users of telecommuting, vanpools, and carpools in the nation.[32] In addition, the cities of Portland and Salem have proved extremely active in a number of greenhouse gas reduction initiatives, working closely with state programs in several cases.[33] Furthermore, the state has been active in public education on climate change and state efforts to reduce greenhouse gases, ranging from community forums across the state to production and distribution of a video, *Generation to Generation: The Story of Climate Change in Oregon.*[34]

However, the overall impact of Oregon's energy-siting legislation has not been as great as anticipated, at least as of late 2003. An initial flurry of interest in new facility siting followed the enactment of the legislation but stalled in 2001 and 2002. After the short-term alarms about the adequacy of electricity-generating capacity to meet demand, interest in developing new capacity has slowed in many regions, including the Pacific Northwest. This appears to be attributable to the availability of substantial new capacity during 2001 and 2002, most of it through fossil-fuel plants sited under aggressive efforts by some states to expand capacity and increased operational performance of the nation's nuclear power plants, which continue to provide more than one-fifth of the nation's total electricity. In turn, the national recession has dampened demand for electricity. As a result, some 5,500 MW of planned capacity expansion around the nation was cancelled in 2002, up from 1,300 MW in 2001.[35]

Energy analysts concur, however, that this transition may be short-lived in the event that economic growth rebounds, demand for cleaner energy intensifies, and aging facilities that produce substantial quantities of electricity from coal, uranium, and water face retirement. In Oregon, most operational dams are between thirty and sixty years old and are unlikely to be replaced; the state also no longer has nuclear-generating capacity and is unlikely to be successful in importing electricity in the future from California. Consequently, the impact of its carbon dioxide standards will best be tested in the longer term, in the event that new electricity-generating capacity is needed in Oregon and the Pacific Northwest.

Linking Information Disclosure with Greenhouse Gas Reductions: Wisconsin's Mandatory CO_2 Reporting and Voluntary Initiatives

Most developed nations, including many that have ratified the Kyoto Protocol, are only beginning to measure their greenhouse gas emissions with any level of precision. A variety of methods exist to estimate releases and are used in current planning for Kyoto implementation. Carbon dioxide releases, for example, tend to be calculated on the basis of reported information on fuel composition and consumption rates, with different fossil fuels generating different emission levels. Methane and other greenhouse gases follow a somewhat similar pattern, with estimates based on volume of certain kinds of activity—for example, numbers of acres planted with a particular crop such as rice. This approach to data gathering is replete with uncertainties, however, particularly given the ubiquity of sources in any polity that generates greenhouse gases. Some analysts have become increasingly concerned that any international regime designed to credit and trade greenhouse gas emissions may founder owing to the lack of more reliable, source-specific data as well as gaping inconsistencies in the methods and analytical capacities of various jurisdictions. Since Rio, countries have begun to report their aggregate emission estimates, but as the policy analyst David Victor notes, "little of the data is actually useful. Reports are incomplete and lack the uniformity necessary to make them comparable." Moreover, Victor laments, "only skeletal international capacity exists to analyze reported data, and major governments seem content to keep that capacity weak."[36]

The United States has made some significant inroads in this area. The 1992 Energy Policy Act required all electricity-generating utilities to provide annual reports to the federal government on their carbon dioxide emissions. Many states have subsequently developed detailed inventories of their own emissions, often supported by federal grants, developing a general profile of their emission sources. This work has, in many instances, helped direct states in pursuing particularly promising reduction opportunities. The State of Wisconsin has gone a step further, however, mandating public disclosure of annual carbon dioxide releases for all large sources as part of its overall system of emission reporting. Wisconsin is the only North American government that has established such extensive reporting, and its emissions profile may be more detailed than those of most developed nations.

The idea of disclosing potentially sensitive information to state governments and the general citizenry is not new in relation to greenhouse gases. The political scientists William T. Gormley Jr. and David L. Weimer have analyzed a wide range of efforts to develop measures of organizational performance that can be shared with the public in various sectors of public policy.[37] Such efforts have expanded and intensified in recent years, but build on historical models. For example, Horace Mann advocated standardized tests to measure the performance of public school pupils as early as the 1840s, and Florence Nightingale pushed to develop "report cards" on hospital performance in the Civil War era.

Development of "performance measures" has long been difficult in environmental policy owing, in part, to the technical difficulties of measuring varied sources of environmental contamination. But both federal and state governments have moved actively into this arena during the past two decades, perhaps most notably through the Toxics Release Inventory. Modeled closely after early state programs in Maryland and New Jersey, this program was established on a national basis in 1986 when it was added to the reauthorization of the Superfund program for cleanup of abandoned hazardous-waste dumps. The Toxics Release Inventory required annual disclosure to the U.S. Environmental Protection Agency of several hundred widely used chemicals. Firms using these materials were required to report their releases to air, land, and water. This information was then to be disclosed to the general public on an annual basis. Over time, the program has been expanded to include additional chemicals, now involving more than six hundred separate substances, as well as a much larger body of private and public institutions that release them. This data source has been used extensively to examine the environmental impacts of a wide range of pollution sources. In turn, a range of federal and state programs have been established that make use of these data, including efforts that target reductions of certain chemicals or challenge particular industries to voluntarily reduce their releases.[38] The Toxics Release Inventory has been widely heralded for its capacity both to directly place important environmental information in the public domain and to provide direction for future emission reduction strategies.

This approach has since been extended to other areas of environmental policy. For example, Congress amended federal legislation on safe drinking water in 1996 to require that locally based water authorities "send their customers annual accounts of detectable contaminants in drinking water."[39] This was a direct response to an earlier outbreak of cryptosporidium in Milwaukee's water supply that resulted in the hospitalization of forty-four hundred people. Indeed, much recent effort to develop new policies to address the "next generation" of environmental problems has emphasized development of more-reliable data to measure environmental performance and guide policy decisions.

In contrast, governments have generally been cautious about including greenhouse gases, including carbon dioxide, in those developing metrics. This may be changing, however, at least at the state level. The Wisconsin disclosure program, for example, coincides with efforts by other states, such as Illinois and New Jersey, to use monthly electricity bills to convey general information about the amounts of greenhouse gases—and other emissions—that are released owing to electricity generation in their region. But only the Wisconsin program creates a detailed and longitudinal base of carbon dioxide emissions that has been linked to other policy initiatives.

The Political Path to Carbon Dioxide Disclosure and Its Utilization

The State of Wisconsin framed climate change as a serious environmental problem in the late 1980s and pursued a number of projects in the subsequent decade to study the state's emissions and consider possible mitigation efforts. Much like Oregon, the state had a long history of active engagement on environmental protection and energy issues. For example, it developed a set of aggressive energy efficiency programs in the 1980s and enacted rigorous sulfur dioxide reduction legislation in 1986, well before federal action. These kinds of innovations developed expertise within the Wisconsin Department of Natural Resources (DNR) and other state government units that proved directly transferable to greenhouse gases.

In turn, DNR veteran George Meyer provided strong support for climate change innovation during the critical period from 1993 to 2001 when he served as department secretary. He championed new initiatives and provided support for entrepreneurs based in the department. "We've always been very up front that we were concerned with climate change, that we were looking at carbon dioxide and other greenhouse gases," recalls a senior DNR official who has been active in climate change throughout the past decade. "Meyer was always good on this issue, and there have been a lot of people in this state taking this seriously," explains a senior member of one of the state's major environmental groups. "I think everyone realized that the state was taking this seriously back in 1997, when Tommy Thompson [the Republican governor from 1991 to 2001, who has subsequently served as secretary of health and human services in the Bush administration] made a Governor's Proclamation on Energy. He said that the global climate change debate was over, and we needed to get serious about it."

This political receptivity to explicit labeling of climate change made the incremental adjustment of mandatory reporting to include carbon dioxide remarkably easy and noncontroversial. The Wisconsin Air Contaminant Emission Inventory already included release requirements that were linked primarily to the federal Toxics Release Inventory. Department of Natural Resources staff advanced the idea of adding carbon dioxide to the list of 546 more-conventional contaminants during a period in which the state inventory was being updated. The Wisconsin Natural Resources Board reviewed the proposal and approved it in May 1993.[40] "There was little discussion or debate over this," recalls one DNR staff member. "We were starting to think about climate change in a number of ways; this seemed a logical step."

Under the new provision, all Wisconsin-based facilities that release more than 100,000 tons of carbon dioxide annually are required to report the extent of these emissions. This includes all electrical utilities and most large industrial boilers in operation in the state. But dozens of sources that fall well below the threshold voluntarily report their carbon dioxide emissions annually, providing the state with a detailed, multiyear profile of a wide range of carbon dioxide sources (see table 3-1). In 2000, for example, 183 state sources reported their carbon dioxide emissions, representing approximately three-quarters of the emissions generated outside the transportation, residential, and agricultural sectors. This includes all the electric utilities in the state, most major industries, and a diverse mixture of smaller sources such as the City of Kenosha Water Utility, the Kraft Pizza Company, the Wisconsin Air National Guard, and various hospitals

Table 3-1. *Wisconsin Carbon Dioxide Emissions under Mandatory Disclosure Program, by Sector, 2000*

Industrial category (SIC code)	Reporting units	Emissions (tons)
Nonmetal minerals (14)	21	6,993
Heavy construction (16)	3	5,165
Food and kindred products (20)	18	99,071
Lumber and wood products (24)	6	3,190
Furniture and fixtures (25)	2	1,400
Paper and allied products (28)	26	7,494,673
Printing (27)	5	1,263
Chemicals and allied products (28)	1	20
Petroleum and coal products (29)	4	4,957
Rubber and miscellaneous plastic products (30)	4	4,893
Stone, clay, and glass products (32)	4	316,118
Primary metal industries (33)	8	21,624
Fabricated metal products (34)	13	19,858
Industrial machinery and equipment (35)	15	315,165
Electric and other electronic equipment (36)	4	22,313
Transportation equipment (37)	7	6,499
Instruments and related products (38)	1	775
Miscellaneous manufacturing industries (39)	1	102
Air transportation (45)	1	622
Electric, gas, and sanitary services (49)	30	42,072,833
Wholesale trade—nondurable goods (51)	2	2,499
Insurance carriers (63)	1	14
Personal services (72)	2	1,357
Health services (80)	8	9,448
Educational services (82)	1	54
Executive, legislative, and general (91)	1	1,011
Justice, public order, and safety (92)	1	2,250
National security (97)	1	51
Other	1	2
Total	192	50,414,110

Source: Data available upon request from the Wisconsin Department of Natural Resources.

and medical centers. The number of reporting entities has increased each year since the program began.

State officials have not attempted any major dissemination of this information to the public, although it is made available to any citizen or organization that requests it. They have thus avoided a tactic used in other information disclosure areas that attempts to draw substantial publicity to large emission sources and influence reductions through a "shock-and-shame" effect. These efforts often lead to listing of the "top ten" or "dirty dozen" emission leaders in media accounts, with the intent of triggering reduction efforts through a combination of public outrage and corporate embarrassment.[41]

Registering Reductions

Rather than resort to shock-and-shame tactics, the state attempted, instead, to use the reporting process alongside development of a multi-faceted approach to climate change, many elements of which could make use of the carbon dioxide disclosure data. The very development of the reporting system coincided with the creation of a Wisconsin Climate Change Committee by DNR Secretary Meyer. This body reviewed the substantial analytical work concerning greenhouse gases and climate change in Wisconsin that had already been completed by the DNR and outlined a series of policy recommendations in a May 1998 report.[42] One of these "proposed actions" involved the development of a system to ensure "credit for early emission reductions" of greenhouse gases: "A significant deterrent to industry adoption of energy efficiency measures and the shift to 'cleaner' fuels as a greenhouse gas reduction strategy is the concern that those who take early actions to reduce emissions would be penalized if a future regulatory program required them to make emissions reductions over and above those they had already made."[43]

This recommendation was translated into a proposal to develop a registry in which any Wisconsin firm could report its reductions of carbon dioxide or any other greenhouse gas. Such reductions would be registered by the DNR with the intent of obtaining credit for reduction in any future federal or state greenhouse gas reduction program. The idea of a registry, of course, was not unique to Wisconsin. The U.S. Department of Energy established a voluntary reduction registry in the early 1990s through the 1605(b) program under the 1992 Energy Policy Act; and other states, such as New Hampshire and California, have developed registries of their own. However, Wisconsin's registry is unique in that it builds on considerable mandatory reporting experience for carbon dioxide and has several design features intended to maximize participation. The legislation authorizing the registry was developed by DNR staff and approved with bipartisan support in the legislature. Governor Thompson signed the bill into law in May 2000, and supportive regulations drafted by the DNR were adopted in 2002 by the Wisconsin Natural Resources Board. This made the registry operational in early 2003.

The creation of the Wisconsin registry introduces important questions. Can voluntary programs achieve meaningful reductions of greenhouse gases, or are more coercive policies necessary? Can a tool such as a registry attract a significant number of applicants and provide an incentive for early reductions, given the opportunity to secure future credit? Thus far, Wisconsin officials are hopeful that several unique features will ensure a high level of engagement. First, the state's considerable experience with carbon dioxide reporting is unique and could indeed simplify the analytical process needed to quantify reductions. A large number of Wisconsin firms already have multiple years of data on carbon dioxide releases on record with the state, some of which indicate reductions that presumably could be entered into the registry. "The link with the emissions inventory is the real hook here, compared to other registries," notes one DNR official. "Firms are already reporting their emissions; we already have all of that data. So this gives us a huge advantage over other registries."

Second, the Wisconsin registry allows for an unusually wide range of air contaminant reductions to be included. Whereas 1605(b) and other early registries have focused solely on carbon dioxide or greenhouse gases, Wisconsin includes all greenhouse gases as well as a large number of other contaminants, including mercury, volatile organic compounds, nitrogen oxides, and fine particulate matter, among others. Entities are allowed to submit reductions for one or more of these contaminants, and state officials believe that this opportunity to register multiple reductions simultaneously will heighten participation. Some of these contaminants are receiving growing attention in the state. In particular, there has been growing concern in Wisconsin about concentrations of mercury levels and their impact on human health. This has triggered increasingly active discussion in the legislature about requiring major new mercury reductions as the state fashions its own version of multipollutant air pollution legislation. Carbon dioxide may also be included in such legislation in some form, making it possible that early involvement in the registry could render credits for early action to reduce emission sources. At the same time, other contaminants such as volatile organic compounds are of interest because of the offset requirements for new sources in ozone nonattainment areas, which is a growing concern for Wisconsin and other midwestern states.

Third, Wisconsin has a well-established tradition of securing credit for steps to reduce air emissions. The state passed major acid-rain legislation in 1986 and influenced the subsequent federal legislation that passed four years later.[44] Wisconsin required public utilities to reduce their sulfur dioxide emissions to 50 percent below 1980 levels by 1994 and assured firms that their reductions would be credited under any future federal law. Most of the state's largest greenhouse gas sources "buy into the registry idea because we came through for them on acid rain," explains

Meyer. "They had to go ahead of other states because of our legislation, but that made it much easier for them to comply with the 1990 law. We said we would make sure they were grandfathered, and they were. They remember that, so they trust us when we say we will make sure their early reductions of greenhouse gases are protected as well. That's what makes this inventory so promising."[45]

The Limitations of Voluntary Strategies

The emphasis on a voluntary strategy and the design of the inventory reflect considerable promise but also serve to underscore the potential limitations of near-term Wisconsin use of its data from the mandatory disclosure program. Perhaps the biggest issue is whether firms will in fact participate. This reflects the "field-of-dreams" qualities of such programs, whereby, given the absence of any state coercion or formal incentive to participate, each firm decides on its own whether it will engage. In turn, DNR staff remain concerned that the process for registering data is considerably less rigorous than the one they had originally proposed, a fact that could raise significant questions of registry legitimacy down the road.

Originally, DNR staff recommended that any emission reductions included in the registry require formal verification, either by state officials or by a third party. But firms likely to participate in the registry balked at having to assume the costs for verification. At the same time, the state proved reluctant to hire additional staff to handle these functions, especially given the budget deficits of approximately $1 billion a year that Wisconsin was facing at the very time the registry was created. "The DNR just doesn't have the resources to invest in double-checking every entry," according to John Mooney, a member of the regional EPA office in Chicago.[46] As a compromise, the state intends to allow for third-party verification, which would be formally noted in the emission reduction listing within the registry. "But there are real credibility issues with the lack of verifiability," acknowledges one state official. "The big trick here has been to strike a balance between something that is credible and workable, especially given our staff and budget limitations."

The state also faced an unexpected setback in a major regulatory innovation initiative that promised to generate substantial reductions of greenhouse gases and conventional pollutants. These reductions could have been measured with precision and been prominent candidates for inclusion in the registry. Through the 1990s, Meyer and the DNR actively explored alternatives to traditional regulatory approaches that might offer complying firms greater flexibility in exchange for overall environmental contamination reductions that exceeded existing standards. As Meyer noted in 1999, "The regulatory system may have reached the limits of its effectiveness and could benefit from more adaptive approaches."[47] In response, the state developed active, collaborative relations with environmental policy officials in the Netherlands and Germany and began to explore new policies that might build on celebrated innovations in those nations.[48]

Given this interest, Wisconsin was a natural participant when the Clinton administration endorsed regulatory innovation under the umbrella of its larger efforts to "reinvent government" in the mid-1990s. In the environmental arena, one of its flagship initiatives was Project XL (Excellence and Leadership), which was launched in 1995. Under this effort, according to President Clinton and Vice President Gore, responsible companies were "given the flexibility to develop alternative strategies that will replace current regulatory requirements, while producing even greater environmental benefits."[49] Participation in the project involved firms anxious to take the lead in exploring regulatory innovation. In some instances, such as Wisconsin, states actively encouraged their involvement and enacted legislation authorizing pilot projects. According to the political scientist Graham Wilson, "[the] DNR took the lead in arguing for a new approach to regulation in the state."[50] In 1997 the legislature enacted and Governor Thompson signed into law the Comprehensive Environmental Agreement Act, authorizing up to ten pilot agreements between Wisconsin firms and government agencies over a five-year period.[51]

This put the full weight of state government behind firms eager to engage with the federal government in the XL process and created the possibility of meaningful three-way collaboration between the private sector and its counterparts in the federal and state governments. The DNR actively encouraged major Wisconsin firms to participate. Six firms submitted applications, the most notable involving the Wisconsin Electric Power Company and all of its generating plants in Wisconsin and Michigan's Upper Peninsula. This company is Wisconsin's largest corporate generator of carbon dioxide and is a major source of numerous conventional pollutants, reflecting its status as the dominant electricity provider for many sections of the state.

Wisconsin entered into a formal memorandum of agreement with EPA on such cooperative projects in 1999. A major delay in these negotiations was created by EPA resistance to DNR efforts to draw directly on the Netherlands' system of regulatory covenants to provide an umbrella over a number of emission sources. This approach emphasizes overall environmental impacts rather than exact compliance with standards for each separate point source of pollution. In the memorandum, the DNR was forced to abandon this idea and work within the more conventional American framework. The negotiations with Wisconsin Electric Power Company then proceeded and tested the extent to which regulatory flexibility could be exchanged for "superior environmental performance." For the power company, a significant attraction was the possibility of "streamlined or reduced administrative procedures for monitoring and data reporting requirements, as well as permit streamlining and an expedited approval for certain kinds of permits."[52] For the DNR, a significant attraction was operational changes that could lead to reductions of conventional pollutants well below current permitted levels and reduction of carbon dioxide emissions well below current operating practice. These included, for example, extensive reuse of coal ash from landfills in generating electricity, offering reduced use of coal, improved land use, and overall emission reductions. "Everyone involved with this saw the potential, including the national office of EPA," recalls Meyer, who was active in the negotiations along with DNR staff. "We put everything on the table, including greenhouse gases. We were looking at a total greenhouse gas reduction from these plants of at least 20 percent over five to seven years."[53]

Like several other prominent XL projects around the nation, the Wisconsin Electric Power Company proposal ultimately collapsed when federal law—or federal agency interpretation of that law—limited proposed flexibility. In particular, EPA officials based in the agency's regional office in Chicago objected to the agreement's perceived inconsistencies with federal New Source Review standards under the 1990 Clean Air Act Amendments. This objection focused on previous New Source Review interpretations that had been applied to comparable facilities operating in the region. The Environmental Protection Agency thus rejected important provisions that offered Wisconsin Electric Power Company greater flexibility in exchange for their commitment to achieve greater overall emission reductions. This conflict brought the Project XL negotiations to a halt in 2000, as the parties could not move beyond this required application of the standards. "The decision by the regional EPA killed it," explains a senior agency official. "We felt we had been responding to the mission of

XL and what all the top people in the Clinton administration told us we could do. This threw all of that into question."

Meyer and the DNR responded to this setback by placing further emphasis on a state legislative proposal to promote a Wisconsin-based system for environmental regulatory innovation. This so-called Green Tier proposal retains many of the concepts borrowed earlier from the Netherlands and Germany, as well as Project XL, but attempts to move beyond pilot projects to standardize them as a set of uniform options under state law. It has many parallels to New Jersey's efforts to link expanded flexibility with advanced reduction of greenhouse gas and conventional emissions. If enacted, the Wisconsin legislation would create a multitier system that would offer the possibility of greater regulatory flexibility to firms that demonstrated superior environmental performance. No state-based firm would be required to participate, and all could indeed remain within their existing set of regulatory relations with the DNR. However, the state would offer separate designation—either tier 1 or tier 2—to firms that achieved specified levels of environmental performance. Each tier, in turn, would offer ascending benefits, including compliance flexibility and public recognition. As the political scientist Graham Wilson has noted, this approach would "regulate environmental high fliers more flexibly, sympathetically and predictably providing an administrative inducement to superior performance."[54]

State policy entrepreneurs have viewed Green Tier as one way to place greenhouse gas issues into a broader review of environmental regulations and thereby provide an opportunity to begin to think—and negotiate—across traditionally narrow categories of regulation. Much like the Project XL proposal, passage and enactment of Green Tier could lead to negotiated reductions that involve a range of emission categories. Such a process could indeed be linked with the voluntary registry, including any greenhouse gas reductions achieved through the negotiations.

The future of Green Tier, however, remains uncertain. Although it passed the Wisconsin Assembly in 2001, it never received extended hearing in the Senate, owing to the consuming focus on the state's fiscal crisis in 2002. It was reintroduced after the 2002 elections, but its status has been clouded by a series of factors. First, senior leadership of the DNR was instrumental in developing Green Tier and its forerunners, playing central entrepreneurial roles that included building support from corporate and environmental leaders. According to Wilson's analysis of Green

Tier, DNR leaders "developed the program, they have been its major supporter, they have recruited allies and supporters and in the face of numerous setbacks, they have persisted."[55] In particular, George Meyer has been a leading champion of Project XL, Green Tier, and the general effort to link greenhouse gas reduction with the larger process of environmental regulation in Wisconsin. However, he was removed as DNR secretary after Thompson joined the Bush administration in 2001, and he retired from the department in April 2002. Some senior DNR staff remain supportive of this approach, as do some key legislative allies, but the earlier base of high-level agency support has clearly weakened.

Second, the state's continuing budgetary distress has resulted in reduction in the size of the DNR and further reduced staff resources available for innovations such as Green Tier or greenhouse gas reduction more generally. Further pressures in 2003 by EPA to accelerate permit approval under Title V of the Clean Air Act have shifted remaining staff from greenhouse gas reduction policy to conventional permitting functions. Third, the Project XL experience gave many participating stakeholders a bad taste of multiyear negotiations that ultimately result in no new policy. It is not clear just how far states can go, politically with federal agencies or constitutionally with federal courts, in crafting their own approaches in regulatory areas with mixed jurisdiction. Potential participants will be mindful of this uncertainty before committing resources to any future deliberations.

Development of Energy Efficiency and Renewable Sources

Consistent with other prime-time states, Wisconsin has also pursued other strategies expected to reduce greenhouse gases. Under 1999 legislation that was focused primarily on increasing the reliability of the long-term electricity supply, Wisconsin created a renewables portfolio standard and a new pool of funds for energy efficiency projects. As was the case in Texas, proponents of these provisions did not expressly emphasize their potential impacts on greenhouse gases. Instead, they stressed that a renewables portfolio standard would most likely deliver supplemental electricity and expanded energy efficiency efforts would dampen demand, both increasing the likelihood of long-term reliability. "At that time, Governor Thompson's rule was that nothing could be done that might harm electricity supply reliability," recalls a senior DNR official. "So there were lots of opportunities to develop new ideas as long as they could be couched with the reliability concern."

The Wisconsin renewables portfolio standard calls for a gradual increase in state use of renewable energy, reaching a level of 2.2 percent of total energy by December 2011. This excludes hydropower sources, which currently supply about 4 percent of total electricity consumed in the state. This provision has clearly spurred new interest in renewable energy, most notably a series of small wind-power projects that have increased the total number of operational wind turbines in Wisconsin from two in 1998 to fifty-three in 2002. In fact, the state has already acknowledged that it is on track to surpass its target well before 2011. This prompted endorsement by both of Thompson's successors as governor, Republican Scott McCallum (2001–02) and Democrat Jim Doyle (elected in November 2002), to raise state renewables portfolio standards significantly through new legislation. Governor Doyle, in fact, has proposed raising the standard to 10 percent by 2013.

The biggest impediments thus far to significant expansion of renewables have been a pair of unanticipated factors. The state has already endured a divisive political controversy over a proposal to construct a major wind farm on agricultural land between West Bend and Milwaukee. Had it been developed, this facility would have approximately doubled total current state capacity for wind power. The proposal was strongly supported by neighboring farmers and by public utilities, which wanted to purchase the electricity. It faced strong public opposition, however, from surrounding communities and was ultimately abandoned in the face of NIMBY-type opposition. Those areas of Wisconsin that are physically best suited for wind power tend to be located in the more populated eastern sections of the state, so siting issues may reemerge if additional proposals for wind power are advanced.

At the same time, Wisconsin would like to import increased volumes of wind power from neighboring states, such as North Dakota and Minnesota, with more favorable physical features for massive wind generation. However, as noted earlier, the fragmentation of interconnections on the national electricity grid make importation of western power into Wisconsin difficult. The western portion of Wisconsin is part of the Mid-Continent Area Power Pool, which includes wind-abundant Minnesota, Iowa, and (most of) the Dakotas, but the Mid-America Interconnected Network covers the eastern two-thirds of Wisconsin (see figure 2-1). Construction of extensive new transmission lines that cut across these pools is the most plausible technical answer, but it raises the likelihood of huge political battles over land use and siting. "We're sitting right next to one of the best wind sources in the country, and we are willing and able to pay for that power as it develops," explains one official who has worked for state government and environmental groups. "But siting new transmission lines across virgin lands is very difficult to talk about politically." Indeed, a loose coalition of local government officials and environmental groups in Wisconsin has blocked proposals for new lines that would ease substantially the import of electricity from western states.

A Common Border and Common Issues: Illinois

Illinois, Wisconsin's neighbor to the south, faces somewhat similar challenges in electricity generation. It is fully included within the Mid-America Interconnected Network and shares a history of electricity transfer—and cross-border air pollution controversies—with Wisconsin. Illinois may be best thought of as a hybrid that cuts across the cells of figure 1-1. On the one hand, it is a heavily industrialized state and has historically provided strong support to its coal-mining industry located downstate. In fact, two statues herald "King Coal" on the grounds of the state capital in Springfield. In December 1998 the state passed anti-Kyoto legislation that included a prohibition against state agency adoption of new rules designed to reduce greenhouse gases. This was not written in quite as draconian a fashion as legislation in some hostile states, and it allowed continued development of voluntary reduction initiatives. But it clearly served as a brake on policy innovation.

At the same time, the state has not shied away from extensive study of climate change, beginning with the development of the Illinois Task Force on Global Climate Change that was formed in 1989 and has continued to operate. The state has also pursued a series of policies, variously labeled, that could foster greenhouse gas reductions. "The great thing about Illinois is that it is a schizophrenic state, and its reaction to climate change demonstrates that," explains a senior member of a leading environmental group. "South of [Interstate Highway] I-80 you have a resource-extraction-based economy and lots of farmers. To the north, you have this big urban and suburban area concerned about things like clean air. They don't match, and policy reflects that." In Illinois, important leadership on climate change has come from within state agencies, most notably the Illinois Department of Natural Resources and the Illinois Environmental Protection Agency, in some instances through collaboration with environmental groups. They have had to operate in a conflicted political context, however, particularly given the constraints imposed by the 1998 legislation.

Consequently, Illinois has moved much more cautiously than other states covered in this chapter and the one that follows. But the state cannot be characterized as either hostile or indifferent upon examination of a series of recent policy initiatives, most of which occurred during 2001. Illinois has designated funding for renewable-energy programs through a four-mill social-benefit charge ($0.004 per kilowatt-hour), ranking second only to California in the total amount of funding expected to be generated for this purpose. The state supplemented this program with $500 million in state revenue bonds dedicated to the development of renewable-energy facilities over the next decade. In addition, Illinois established a renewables portfolio provision that sets a nonbinding "goal" of 5 percent by 2010 and 15 percent by 2020. Legislation is pending in the state Senate and General Assembly to shift these targets from goals into a formal standard. Illinois is one of five states to mandate minimum purchase of energy from renewable sources in government-operated facilities.[56]

The state has also begun to explore ways to address greenhouse gas emissions from electric utilities. Legislation enacted in 1997 mandates disclosure of carbon dioxide emissions from regional utilities in the electricity bill mailed monthly to every consumer in the state. The legislature has continued to examine ways to integrate carbon dioxide reductions with reductions of conventional pollutants, through a series of proposals that resemble the strategies employed in New Hampshire and Massachusetts. The Illinois Department of Natural Resources has considered a number of additional greenhouse gas reduction options and has even engaged in shuttle diplomacy with a sister province in China, examining possible greenhouse gas reduction projects and long-term prospects for emissions trading. Recovery of methane from farming and mining practices is also an ongoing area of exploration in Illinois.

Cutting across Policy Sectors

Illinois is one of a growing number of states pursuing multiple initiatives that affect greenhouse gases, although it clearly lacks the same level of engagement as other prime-time states. Policy entrepreneurs in that state have moved forward on a number of fronts, but they operate within a more restrictive political context than the three states examined earlier in this chapter. New Hampshire, Oregon, and Wisconsin, among others, have been able to frame climate change as an environmental threat that warrants a serious policy response. Each state has fashioned its own particular policies, developing very different tools to attempt to reduce greenhouse gas releases from within its boundaries. All of these have been explicit in labeling their intended goal as greenhouse gas reduction but have been sensitive to their particular economic circumstances. This has enabled entrepreneurs to build solid coalitions to support enactment and implementation.

At the same time, these cases reflect limitations in what states have been able to undertake thus far. Each state has concentrated on select aspects of its total emissions rather than a cross-cutting approach. In turn, each has begun to face limitations in reducing its overall impact in the very sectors in which it has concentrated. New Hampshire has made a concerted effort to reduce carbon dioxide releases from electrical utilities, but it remains a small state that is likely to continue to purchase at least some of its electricity from states and Canadian provinces that lack comparable standards. Oregon's focus on carbon dioxide emissions standards for new power plants has had some significant impact, but the state must contend with broader forces such as slackened demand to construct new electricity generation sources. The constraints facing Wisconsin may be most significant in the long term, raising the question of just how supportive the federal government is prepared to be toward states taking climate change policy initiatives. Will emission reductions incorporated into the Wisconsin registry be recognized and accepted at the federal level? Will the state be able to develop alternative approaches to environmental regulation that incorporate greenhouse gases while passing muster with federal officials determined to secure compliance with conventional requirements? Will it find ways to create the infrastructure, whether siting new wind generators or transmissions lines, to achieve its expanding goals for renewable-energy use?

New Hampshire, Oregon, and Wisconsin do not constitute the end of the road for prime-time states. Some additional states have been able to develop an even more active and comprehensive approach to climate change. Much of the analysis that follows focuses on New Jersey, but it also extends to recent developments in states not formally included in the study, namely California and the five other New England states that have begun to work in concert with New Hampshire.

CHAPTER 4

An Unlikely Front-Runner in Climate Change Policy

From wind energy in Texas to carbon dioxide standards in New Hampshire and Oregon, the elements of a bottom-up American approach to climate change are beginning to take shape. States have clearly positioned themselves to achieve significant reductions in greenhouse gas emissions through the kinds of policies introduced in the previous two chapters. Some states are beginning to make important linkages between policies to leverage greater reductions, as is evident in the primetime cases examined in chapter 3. The connections between energy development and forestry in Oregon and the potential ties between emissions disclosure, reduction registry, and negotiated reductions in Wisconsin illustrate a shift from policies concentrated in a single sector toward a more robust, integrated approach. This shift reflects the realities of greenhouse gas releases, which are not confined to a particular policy sector or jurisdiction. These initiatives reflect a framing of climate change as a serious environmental threat that warrants a serious response with explicit labeling of new policies designed to reduce greenhouse gas releases.

But at least one additional case in the study indicates that states may be prepared to go still further, cutting across multiple sectors in pursuit of a comprehensive strategy to achieve significant greenhouse gas reductions. This case involves the only government in North America that has, in essence, embraced both the Rio Declaration and the Kyoto Protocol and has taken formal steps to ensure implementation. Whereas the United States has withdrawn from the Kyoto agreement, Canada dithered for more than half a decade before deciding to ratify the protocol, but it has not yet outlined serious steps to achieve reduction commitments, and Mexico has not been a participant in negotiations. In contrast, one American state, New Jersey, made a formal pledge in 1998 to achieve reductions by 2005 that would put it in line to reach Kyoto-level emissions by the end of the current decade.

An American state, of course, cannot ratify an international treaty. But this has not prevented New Jersey from conducting itself like a cooperative participant on the international stage. In so doing, the state has been up front about its intent, characterizing climate change as a serious environmental threat but pursuing reductions that make economic sense and have proved politically feasible. New Jersey has specified its reduction goals, developed a strategy that involves all relevant sectors, and secured a broad base of support. It may, in fact, be substantially more prepared to achieve Kyoto-like reductions and work effectively with other governments toward this end than many of the European and other developed nations that have ratified the Kyoto Protocol and have made formal commitments to reduce their emissions.

New Jersey is, in many respects, a surprising candidate for national leadership on climate change policy. For many Americans, New Jersey is commonly known as a home to more abandoned hazardous-waste dumps than any other state, owing, in large measure, to its dense concentration of industries that make extensive use of hazardous chemicals. It is often the butt of jokes about the questionable ethical practices of some of its prominent citizens, whether politicians such as former U.S. senator Robert Torricelli or television's Soprano family.[1] The state has also experienced substantial political turbulence in recent decades, perhaps best reflected in a series of acrimonious and highly partisan shifts in the control of the office of governor.

But these vicissitudes have not prevented New Jersey from framing climate change policy as a necessary response to a serious environmental threat. The governorship of Christine Todd Whitman, from 1994 to 2001, has been the most significant in terms of policy development and implementation on greenhouse gas reduction, building on a modest foundation established in the late 1980s. This is somewhat ironic for a pair of reasons. First, Whitman's overall environmental record was highly controversial, including a series of major state budget cuts that resulted in significant disarray in key agencies such as environmental protection. Her administration was also widely criticized by environmental groups for allegedly weak enforcement of a number of state and federal environmental laws. Second, Whitman carried her active New Jersey record on climate change policy into the Bush administration. It remains unclear what policies she recommended before closing ranks behind the president's decision to pursue a far more modest national strategy than what was already being implemented in states such as New Jersey and Texas. As one former aide notes, "Clearly, something happened on the train ride from Trenton to Washington."

Before assuming federal office, Whitman presided over a significant set of new policy commitments and initiatives designed to reduce greenhouse gases and give New Jersey a national and international presence on this issue. These commitments were explicitly labeled as intended to reduce greenhouse gases, reflected in a statement made by Whitman before her departure for Washington:

> New Jersey has set an ambitious goal to not only curb greenhouse gas emissions, but to reduce them. Our target for 2005 is a 3.5 percent reduction below the 1990 levels. It's a goal to which we are firmly committed, and one that we're relying on the support of private businesses to help us meet. In addition, working in partnership with The Netherlands, we are developing prototypes to carry out an emissions credit trading pilot.
>
> The fact is that climate change associated with greenhouse gases has an effect on every aspect of our daily lives. The environmental and economic benefits that stem from controlling greenhouse gases are enormous. Whether it's maintaining our vibrant shore economy, preserving open space or expanding development opportunities in our urban core, the joint efforts we are undertaking with our business leaders and others are helping to make New Jersey an even better place to live, work and raise a family.[2]

New Jersey followed a pattern of policy formation similar to that in the other prime-time states. Agency-based entrepreneurs played central roles in developing the case for engagement on climate change, devising policies to produce desired reductions and winning over key coalition partners, including elected officials such as Whitman. A pivotal entrepreneur was Whitman's appointee as commissioner of the Department of Environmental Protection (DEP), who passionately gave climate change center stage during much of his eight years in office. The policy development process entailed both broad pledges—including the emission reduction target embraced in Whitman's statement—and specific policy tools designed to enable the state to meet that target.

The Political Path toward a Cross-Cutting Approach to Climate Change

New Jersey launched an array of innovative environmental protection programs in the 1980s and the early 1990s, many of them in reaction to the discovery of extensive toxic contamination. The state's early engagement on hazardous waste and disclosure of toxic emission releases provided models that were ultimately adopted for the entire nation by the federal government. Moreover, the state experimented with a number of efforts to make traditional regulatory programs more effective. These included a pioneering initiative to end the shifting of pollutants across the environmental media of air, land, and water by integrating the issuance of medium-specific permits.[3]

By contrast, the state moved much more cautiously into the area of climate change. Toward the end of his two terms as governor (1982 to 1990), Thomas Kean issued an executive order in 1989 that called upon all units of New Jersey government to take the lead in reducing greenhouse gases.[4] This was consistent with early efforts in other states, which, though largely symbolic, set the stage for subsequent actions. "It was quite progressive for its time, but it never went anywhere," recalls a veteran state official with long-standing engagement on greenhouse gas issues. "When the [Democratic governor James] Florio administration came in during the early nineties, they seemed kind of embarrassed by it. They wanted a strong environmental reputation, although they never acted on the executive order. But they never rescinded it, either."

Meanwhile, state officials in the DEP and the New Jersey Bureau of Public Utilities (BPU) aggressively pursued federal grants to allow them to begin to study various aspects of climate change and energy efficiency. Like other states, New Jersey began to inventory its emissions, drawing on data from the state's energy master plan as well as other sources.[5] They also began to consider options for possible next steps in climate change policy. "For eight to ten years, a number of us snuck around, got some federal money, and proposed different things on greenhouse gases," recalls another DEP official. "But then a commissioner came along who said that climate change was our fight, and things really took off."

That commissioner was a somewhat unlikely candidate to emerge as one of the strongest advocates in the nation for aggressive state action on climate change: Robert C. Shinn Jr. Before assignment as DEP commissioner under Governor Whitman, Shinn had served for twenty-six years as an elected official, initially as mayor of Hainesport and later in the New Jersey Assembly. His strongest environmental interests involved land use, and upon taking office in 1994 he had only passing familiarity with climate change and greenhouse gases. He became a lightning rod for numerous critics who contended that environmental stewardship had been weakened in the Whitman years. A comprehensive assault on greenhouse gases seemed highly unlikely under his leadership.

Sea-Level Rise and Climate Change

As he found himself immersed in issues that appeared to have a linkage with climate change, Shinn began to be persuaded of the need for significant state action. In particular, rising sea levels became an increasingly controversial issue during his tenure as commissioner. Like many coastal states, New Jersey has substantial residential and commercial development within close range of its extensive ocean coast. But the state may face particularly severe threats from sea-level rise. The New Jersey Geological Survey has concluded that the rise of sea level along New Jersey coasts over the past eighty years is "about twice the world rate." The world average for sea-level rise during this period was about 0.04 to 0.08 inches a year. But in the Atlantic City area, reflective of statewide trends, that increase has been 0.15 inches a year. This unusually high rate of increase "may be the result of land subsidence along the coast due to sediment compaction."[6] As a result, New Jersey anticipates serious threats in coming decades to its long and narrow barrier islands, low-lying salt marshes, tidal flats, coastal real estate, and recreational beaches.

Such threats are not unique to New Jersey, of course. Other states that appear to face even greater threats from sea-level rise, such as Louisiana and Florida, have responded with indifference. But Shinn began to take this connection seriously in 1994, during a trip to the Netherlands to meet with its senior environmental and energy officials. New Jersey and the Netherlands have a long history of collaborative work on environmental issues, stretching back to the mid-1980s. Their partnership has continued in recent years, exploring common strategies to reduce green-

house gases, given their strong similarities in physical features and industrial and population densities. "For the Netherlands, climate has been the
number one issue, in part because one-third of their country is potentially
under water," Shinn notes. "They really take this linkage seriously, and
you could see so many parallels to the New Jersey situation."[7]

Shinn returned from the trip with a new interest in climate change and
began to interact more frequently with DEP staff who understood the
issue and had considered possible policy options. This coincided with
New Jersey's development of a greenhouse gas emissions inventory and
action plan, sponsored through EPA grants. Shinn began to take a more
public stand on climate change in 1996, linking it directly to growing
concerns about sea-level rise. In particular, a 1996 report by Rutgers University had recommended adaptation strategies, including abandonment
of coastal islands, in response to rising sea levels. "We were having these
hearings on the future of these islands, and this huge debate erupted over
what to do," recalls Shinn.

> So I decided to raise the issue: Doesn't it make sense to reduce our
> CO_2 emissions? If we started a program, hopefully we could get
> other states involved and slow the increase. I was really stunned by
> the positive response from the audience. It ended the gridlock.
> Clearly, we were not prepared to abandon the islands, and reducing
> CO_2 wasn't the only answer. But it began to give us a constructive
> way to begin to think about this issue.[8]

Shinn decided to push his New Jersey experience to the national level
through his leadership role in the Environmental Council of the States. As
states were meeting frequently to negotiate a cooperative agreement on
interstate management of ozone, they began to recognize linkages with
other issues, including greenhouse gas emissions.

In 1996 a proposal surfaced for a workshop on greenhouse gases at the
next ECOS meeting. "There was a deafening silence in the room when the
subject arose, especially on the question of whether anyone would agree
to lead the session," recalls a senior DEP official who worked closely with
Shinn on climate change issues. "But Shinn agreed to do it." As a result,
Shinn intensified his study of greenhouse gases and began to see genuine
reduction opportunities. He also hoped to use the ECOS session to trigger multistate collaboration. "I was almost burned at the stake by some
ECOS states for even talking about this," Shinn recalls. "I was supposed
to cochair a two-and-a-half-hour session with David Gardiner [executive

director of President Clinton's White House Climate Change Task Force]. But he didn't show, so I ran the session. I really expected a lot of states to come forward, and some did. But many just didn't see the benefits, especially the possibility of linking this with other issues."[9] Shinn left these meetings determined to continue to push climate change onto the ECOS agenda, leading the charge for the 1998 meetings in Madison that he helped organize with allies from several other states. But, perhaps more important, he expanded his work with staff to develop a climate change strategy for New Jersey.

Toward an Administrative Order and Statewide Reduction Goal

State officials were examining different policy sectors and evaluating the potential impact of various strategies. In particular, energy efficiency received intensive scrutiny, since it seemed to offer numerous opportunities to achieve reduction and promote economic development. But the DEP shied away from launching a few specific programs, following Shinn's sense that the state first needed to make some overall commitment on greenhouse gas reduction. "We knew we needed a goal and felt there was an opportunity to set some statewide target," he recalls. "By 1998, we had a pretty good sense of our emissions due to the inventory. We didn't want to call this a Kyoto plan, so we took half of 7 percent for 2005."[10] A senior staff official confirms that "Commissioner Shinn looked at Kyoto and thought that was just too big and too far off. So he cut the timetable and level basically in half, and we thought we could do that."

Consequently, the DEP drafted a statement that would "get New Jersey halfway to Kyoto" by the middle of the decade. Whereas the United States had pledged a 7 percent reduction below 1990 levels by 2012, New Jersey endorsed a 3.5 percent reduction below its 1990 levels by 2005. Of course, an agreement between a state agency head and senior agency staff did not translate into new legislation, opening the question of how New Jersey was to proceed to meet that goal. In particular, DEP officials were concerned about potential opposition. This initiative was being explored in the early months of 1998, at the very time that a number of states were registering formal opposition to what the Clinton administration had negotiated at Kyoto in late 1997. The New Jersey legislature was also exploring during this period a variety of ways to weaken a number of key environmental laws. "We really did not want to get the legislature involved at this point," notes one senior DEP staff member. "We thought this could open up a messy debate there, and we had not been working with them closely on this issue."

Instead, Shinn and the DEP began to explore the possibility of issuing an administrative order embracing the reduction as state policy. The New Jersey constitution vests unusually strong powers in the executive branch, including executive departments. Cabinet members such as DEP commissioners can issue administrative orders, much as a governor can issue an executive order. (New Jersey's first policy statement on climate change was Governor Kean's 1989 executive order.) However, state department heads use this power cautiously and act only with the approval of the governor. Consequently, any administrative order would require Shinn to gain the support of Governor Whitman. "The commissioner took the idea to the governor, and she was opposed at first," recalls one senior DEP official who was active in this process. "But he kept pushing it, stressing both the environmental and economic advantages, and she was finally convinced. She eventually became an advocate, talking it up in public settings and appearing when new steps would be announced."

Whitman's approval enabled Shinn to put New Jersey on record behind the proposed greenhouse gas reductions. Administrative Order 1998-09 was signed by Shinn on March 17, 1998. The order established as policy goals for the DEP "to support and advocate for legislation both state and federal, which has as its goal the reduction of greenhouse gases, and the protection of our coast line from sea level rise, which is one direct result of climate change."[11] In addition, the order included provisions for the DEP to work cooperatively with other parties to find ways to achieve the reduction goals. This included such state entities as the BPU, the Department of Transportation, New Jersey companies, and other countries, such as the Netherlands.

Developing a Multisector Strategy: The State Action Plan

The administrative order, however, was only the beginning of a process to decide how to achieve the emission reduction goal. This task was delegated to the DEP's Greenhouse Gas Workgroup, which was charged with "[focusing] its efforts on the policies set forth herein and [coordinating] the resources of the department to effectuate the goals outlined above."[12] The workgroup consisted of a set of state officials with diverse training and specializations, including the department's leading experts on climate change, drawing heavily from both the DEP and the BPU. It was centered in the Office of Innovative Technology and Market Development in the Division of Science, Research, and Technology. This placement made it possible for the workgroup both to tap into technical expertise in the technology office and to remove climate change from dominance by a more traditional, potentially turf-protective unit of the DEP. "This was a safer place to put this," one DEP official has noted. "It didn't issue any permits or do any enforcement. We were free to think about how to proceed."

Despite significant state budget cuts and staff reductions, Commissioner Shinn committed departmental resources that were coupled with extensive federal grants to create a large team of professionals. Staff members possessed expertise in areas such as energy efficiency, utility regulation, air-quality management, solid-waste management, climate science, and environmental education. "A number of us came out of areas such as solid-waste reduction, pollution prevention, and life-cycle analysis," notes one staff member. "This gave us a fairly broad understanding of how climate change policy might be developed. A lot of us saw it as similar to solid-waste programs, where you had to involve the public and cut across traditional sectors if you were going to be successful."

This team had a unique opportunity to fashion a policy response that could achieve the reduction goals set forth in the administrative order. "This was Bob Shinn's issue, but he gave us free rein to develop the program," one senior member of the team explains.

> We had his full support, and he was really into this. I'd get calls on Saturday at six o'clock in the morning. He'd be at the Home Depot, having seen some new technology that sparked an idea. He wanted me to know about it and do something with it. He knew the business side and saw the need to make New Jersey competitive in the developing energy market. We tried to pull all the pieces together.

Staff at the DEP and the BPU who have worked on this project concur that they had unusual freedom to think creatively and to cut across traditionally fragmented areas. "This is actually the most fun I've had at the DEP," recalls a veteran official who has been active in greenhouse gas policy development. "But it also takes more work—a lot more, if you ask my staff—to do it this way than to simply write a rule that requires something to be done. And not that rulewriting is easy—or that compliance monitoring is easy."

The workgroup not only tried to develop policy proposals but also began to build a coalition of supportive constituents. This included consultation with industries as well as environmental groups that built toward the launch of a comprehensive plan to attempt to hit the reduction goal. It was often a difficult process. "Shinn was a great salesman for this, and it was clear he was a believer in reducing greenhouse gases," explains a director of one of the state's largest environmental groups. "But most of the main environmental groups in New Jersey aren't focused on energy or greenhouse gases. In fact, many of them had been very nervous about any reduction of command-and-control approaches for traditional pollutants and had spent a lot of time trashing Shinn on other issues. So it was hard to try to pull this together." At the same time, a number of New Jersey industries and state agencies showed reluctance to fully engage in this process.

Nonetheless, the DEP released the New Jersey Sustainability Greenhouse Gas Action Plan on April 17, 2000. It was formally endorsed by Governor Whitman and launched at a ceremony that included significant representation from the environmental community, industry, electric utilities, and various state agencies. The plan reaffirmed the statewide greenhouse gas reduction goal and offered a sobering analysis of the amount of reduction that would have to be achieved. It concluded that under a "business-as-usual" scenario, New Jersey's greenhouse gas emissions would be 15 to 16 percent above 1990 levels by 2005, consistent with national trends. Consequently, attainment of an emissions level 3.5 percent below 1990 levels by 2005 would require aggregate reductions of nearly 20 percent over a five-year period.

The Action Plan proposed a multifaceted approach that blended a series of existing efforts with a set of new initiatives. Core strategies covered seven distinct areas: energy conservation, innovative technologies, pollution prevention, waste management and recycling, natural resources and land use, research, and outreach and education. Introduction of new technologies, many designed to increase energy efficiency, was strongly emphasized in several of these areas. Overall, a combination of energy conservation and introduction of innovative technologies in commercial and residential buildings, as well as industrial facilities, was expected to achieve almost three-quarters of the total reductions sought by 2005 under the Action Plan (see table 4-1).

Proponents contended that this effort would allow the state to realize potential environmental benefits and economic development advantages by taking a lead role in this area. These kinds of technologies, ranging from geothermal heating and cooling systems to energy-efficient lighting

Table 4-1. *Proposed Greenhouse Gas Emissions Reductions under New Jersey Sustainability Greenhouse Gas Action Plan, by Market Sector*[a]

Million metric tons of carbon equivalent

Market sector	1	2	3	4[b]	5[c]
Energy conservation					
Residential buildings	25.8	24.3	1.3	6.0	6
Commercial buildings	26.7	18.5	4.5	16.9	22
Industrial	25.6	22.9	0.4	1.6	2
Transportation	46.2	44.0	1.5	3.3	7
Innovative technologies					
Residential buildings	n.a.	n.a.	0.2	0.8	1
Commercial buildings	n.a.	n.a.	3.7	13.9	18
Industrial	n.a.	n.a.	2.3	9.0	11
Transportation	n.a.	n.a.	0.7	1.5	4
Other					
Pollution prevention	5.4	4.6	0.8	14.8	4
Waste management	19.8	15.3	4.5	22.7	22
Natural resources	1.7	1.2	0.5	29.4	3
Total	151.2	130.8	20.4	13.5	100

Source: New Jersey Department of Environmental Protection, *New Jersey Sustainability Greenhouse Gas Action Plan* (Trenton, 2002), p. 11.

a. Column headings are as follows: 1—Projected 2005 emissions with business as usual; 2—Projected 2005 emissions under proposed New Jersey Sustainability Greenhouse Gas Action Plan; 3—Emissions reductions in 2005 under proposed Action Plan; 4—Percentage reduction in 2005 under proposed Action Plan; 5—Percentage of total Action Plan reductions.

b. Column 3 divided by column 1.

c. Column 3 divided by column 3 total (20.4).

and vehicles, "all make economic sense and can help achieve our goal of reducing greenhouse gases," Shinn and DEP official Matt Polsky have written. "Many of these technologies are developed and manufactured in New Jersey, thereby helping our economy as well as the environment."[13] Indeed, during his term as commissioner, which ended in January 2002, Shinn probably emphasized this aspect of greenhouse gas reduction more than any other. "The real mission in CO_2 control is to help people find out how to use energy more efficiently," he explains. "This makes us go out and sell the program, which is a new experience for us. We're used to telling people what to do."[14]

Covenants of Sustainability

One core element of the Action Plan that allowed for active consideration of energy efficiency and new technology involved the creation of "covenants of sustainability" that would commit signatories to the statewide greenhouse gas reduction goal. Such covenant opportunities have been made available to virtually all private, public, and nonprofit institutions in New Jersey, from large corporations to local schools and church congregations. In most instances, these are not legally binding agreements that can result in financial or criminal penalties if pledges are not achieved. However, they do constitute a formal, public commitment to the New Jersey goal and already involve an extremely diverse set of state organizations.

As is evident in other areas of environmental policy in the state, the covenant system is derived from New Jersey's close working relationship with the Netherlands. Under Dutch environmental law, industry frequently pledges to meet broad environmental goals but has considerable flexibility in determining how to achieve these goals.[15] The Netherlands' approach to Kyoto compliance, for example, has relied heavily on creation of a new covenant system whereby "industry undertook to achieve the highest standards in the world within five years."[16] Implementation then occurs through collaboration between individual industrial sectors and governmental agencies with jurisdiction. Much like the Wisconsin experiment that attempted to trade regulatory flexibility for superior environmental performance, environmental policy in the Netherlands routinely follows this pattern. It is also evident, to varying degrees, in other nations of the European Union and offers them considerable flexibility in attempting to honor their commitments to ratify the Kyoto Protocol.

In the New Jersey system, covenant signatories meet with DEP officials and outline a written agreement that specifies their plans for greenhouse gas reductions. Some of these reductions may already have been planned, either as part of an industry modernization program or in conjunction with other voluntary programs to increase energy efficiency.[17] In many instances, the pledged reductions are substantial, well beyond the levels set out in the administrative order. For example, the Johnson and Johnson Company, a pharmaceutical giant based in New Brunswick, has pledged to reduce its carbon dioxide emissions by 18 percent and its nitrous oxide emissions by 21 percent between 1990 and 2005. The firm will use a combination of energy efficiency measures, fuel switching, and load control to achieve these reductions and is attempting to transfer its experience in New Jersey to all of its facilities. "In New Jersey, we will lead and the rest will follow," explains Harry Kaufman, Johnson and Johnson's corporate energy director.[18] L'Oreal USA has, in turn, pledged a 22 percent reduction in its state carbon dioxide emissions during this period, despite an anticipated increase of 60 percent in production, through use of innovative technologies, energy efficiency, pollution prevention, and waste reduction techniques. Public sector entities have also begun to participate. The Naval Air Engineering Station in Lakehurst, for example, has pledged a 34 percent reduction in carbon dioxide emissions between 1990 and 2005, through a range of initiatives.

Some covenants can also add binding provisions, including monetary penalties for noncompliance. This is the case with what is perhaps the most significant greenhouse gas covenant signed in New Jersey to date, part of a consent decree involving a series of air emission issues between the DEP and the state's largest electric utility, the Public Service Enterprise Group. This January 2002 agreement, signed by DEP commissioner Shinn and Thomas R. Smith, the president of the electric utility, includes formal commitments for the utility to reduce its total carbon dioxide emissions from all plants fired by coal, natural gas, or oil that it operates or owns in New Jersey by 15 percent by 2005 from a 1990 baseline. This covenant includes detailed provisions on the exact performance changes, including a limit on carbon dioxide emissions of 1,450 pounds per megawatt-hour (MWh) by December 31, 2005, down from a 1990 baseline of 1,706 pounds per MWh. It also outlines procedures for regular reporting on performance and specifies monetary penalties for any compliance failures.[19] The covenant does not specify how this reduction will be achieved, although use of new technologies and fuel substitution from coal to natural gas are the most likely approaches.

The New Jersey covenant approach has also sought participants that are not conventionally considered as serious contributors to environmental contamination yet may collectively generate significant amounts of greenhouse gases. Higher education is a major undertaking in New Jersey, ranging from elite private universities such as Princeton to a series of state-sponsored colleges. In 2001 the presidents of all fifty-five New Jersey universities and colleges signed the New Jersey Sustainability Covenant. In fact, these institutions have created an organization, the New Jersey Higher Education Partnership for Sustainability, that examines common approaches to greenhouse gas reductions and other initiatives to reduce environmental degradation. Each school has agreed to designate a sustainability coordinator to participate in the partnership's deliberations and a faculty liaison to increase faculty involvement in these issues. "We saw the New Jersey Action Plan as a great umbrella to get involved in these issues," explains the coordinator from one university. "This process has given us more access to our presidents and other university administrators and a chance to demonstrate that we can save them a lot of money by trying these new things." For example, Richard Stockton College, in Pomona, has established an extensive system of geothermal heating and cooling, which has saved the college considerable sums on energy costs and will enable it to meet its covenant pledge.[20] Other institutions have developed detailed profiles of their energy use and greenhouse gas releases in an attempt to develop long-term strategies to ensure continuing reduction of total releases. This effort by New Jersey colleges and universities has been joined at the public elementary and secondary school level as well. A number of public school districts have now signed covenants, and a nonprofit organization, the New Jersey Sustainable Schools Network, is working with individual districts and schools on implementation strategies.

The covenant system also extends to smaller institutions that rarely have any contact with agencies such as the DEP. One of the most enthusiastic supporters of this program has been New Jersey Partners for Environmental Quality, a consortium representing more than six thousand religious congregations that includes members from nine different denominations. The organization, which signed a covenant with the DEP in June 2001, works with individual congregations and denominations to study their own greenhouse gas emissions and to sign reduction covenants of their own.[21] "When we first began to talk about this," notes one member,

> our eyes glazed over; it was just overwhelming for us to have to think about global warming. But we can potentially reach one million of the 3.6 million households in New Jersey, plus our denominations operate all sorts of hospitals, nursing homes, and schools. We have found that, for the most part, our members are completely unaware of things like energy efficiency, much less greenhouse gases. Signing these agreements opens things up, helps these congregations think about long-term possibilities.

Much like their university and public school counterparts, a growing number of New Jersey congregations are participating in the covenant system and exploring possible greenhouse gas reductions.

Links with Electricity Restructuring

At the same time New Jersey has emphasized a more voluntary system of covenants as one cornerstone of its broad strategy, it has also introduced more regulatory mechanisms. Much as in other states, such as Texas, New Hampshire, Wisconsin, and Illinois, electricity restructuring legislation in New Jersey includes provisions that foster expanded use of renewable energy and promote energy efficiency. In many respects, however, New Jersey goes considerably further than these states, both in the numeric levels set in key provisions in the legislation and in the range of initiatives incorporated into a single bill. In fact, New Jersey is one of only three states—joined by Connecticut and Massachusetts—to simultaneously embrace a renewables portfolio standard, establish a social-benefits charge to fund energy efficiency programs, and mandate disclosure to consumers about carbon dioxide emissions released through local electricity generation. These provisions were incorporated into the 1999 New Jersey Electric Discount and Energy Competition Act.

This portion of the state's climate change strategy received less extensive and explicit labeling as a greenhouse gas strategy at the time of its creation than did the covenant system. The predominant purpose of the act was to establish basic rules to guide electricity restructuring. The BPU and the DEP provided key support for the provisions relevant to greenhouse gases and are now playing lead roles in their interpretation and implementation. But they proved somewhat reticent in noting the climate change impacts relative to other benefits, such as energy efficiency and long-term electricity supply reliability. In this regard, they may fit the pattern of cases in cell 2 in figure 1-1, those in which the greenhouse gas ramifications of a particular policy may be played down yet contribute to the climate change strategy of a prime-time state.

Since enactment of the statute, the state has incorporated the legislation into its Sustainability Greenhouse Gas Action Plan, seeing it as vital to the state's greenhouse gas reduction goals because it involves both the generators and the consumers of electricity. Renewables are actively promoted through a portfolio standard that closely resembles programs established in Texas and several other states. New Jersey has only limited prior experience with renewable energy. It also has population density levels and other physical features that may make extensive development of wind and solar power particularly difficult. Nonetheless, the Electric Discount and Energy Competition Act sets a fairly ambitious percentage level for renewables that rises consistently until it reaches a level of 6.5 percent by 2012. It allows New Jersey utilities to achieve the requirements of the state's portfolio standard through a renewable-energy-trading program created by the BPU and the DEP that allows trading of renewable energy across state lines. According to the legislation, "all out-of-state suppliers wishing to do business in New Jersey must comply" with the portfolio standard.[22] This was a particularly significant concern for New Jersey, given its small size and the substantial amount of electricity that moves across state borders in the northeast.

Achievement of these emissions targets may be facilitated in part by a significant new source of state funding to promote renewable energy and energy efficiency. New Jersey's social-benefit charge functions much like those in some other states, including the one in New Hampshire. However, at 26 mills ($0.026) per kilowatt-hour, this "mini carbon tax" is set at a much higher rate than in almost all other states—nearly ten times the rate of such states as Wisconsin and New Hampshire.[23] The social-benefit charge is expected to generate more than $1 billion for new energy initiatives over its first eight years of operation. Under the legislation, 75 percent of the funds are to be allocated "to help buy down the cost of energy efficiency and to transform the market place for energy efficiency." The remainder has been used to create a Class 1 Renewable Energy Fund that will place particular emphasis on wind, solar, and fuel-cell technologies, designed to "buy down the cost of these technologies and assist in market transformation."[24]

The New Jersey BPU launched this program in March 2001, proposing $358 million in expenditures during the first three years of the program. Initial expenditures have concentrated on support for energy efficiency in new construction and residential retrofits as well as audits and training.[25] It also established a process to create a new position, independent statewide administrator, to allocate the funds, rather than leave responsibility for allocation in the hands of utility companies. This reflected a strong concern by the BPU and the DEP that the state needed to make a clean break from the well-established demand-side management programs that had promoted energy conservation in New Jersey and other states but were operated by utilities and tended to retain a narrow focus. "The creation of the [social-benefit charge] and having an independent administrator makes possible options we never could explore under demand-side management," notes the coordinator of greenhouse gas reduction at a major New Jersey university. "Those programs do a lot of good, but smaller institutions and newer technologies never had a chance."

At the same time, those earlier energy conservation programs have not necessarily disappeared since the creation of the social-benefit charge. The 1999 restructuring legislation emphasized that "nothing in this act shall be construed to abolish or change any social program required by statute or board order or rule or regulation to be provided by an electric public utility."[26] This stipulation will serve to protect a series of earlier energy conservation programs and allow for a negotiated transition of earlier demand-side management programs as new social-benefit charge expenditures come into play. "The legislation gives a lot of room to the BPU, which is required to consult with the DEP, to interpret these provisions and set the long-term course," one veteran state official explained. "A number of the [demand-side management] programs may not necessarily continue over time, but we also are developing mechanisms whereby more money will become available to the [social-benefit charge] as utilities pay down their debt under the terms of restructuring. Fortunately, the BPU and the DEP work very well together; both see this as a real driver behind a lot of things we want to do with greenhouse gases."

The final leg of the three-legged greenhouse gas stool included in the restructuring legislation involves environmental disclosure requirements. Unlike the Wisconsin program that mandates individual sources to report carbon dioxide emissions, the New Jersey legislation requires all electricity suppliers, whether or not they are based in the state, to provide detailed information on the bills or contracts of all customers "about the environmental characteristics of the energy purchased." This includes information on emissions, to be reported in pounds per MWh of carbon dioxide, some conventional pollutants, and any other pollutants that the BPU "may determine to pose an environmental or health hazard."[27] This program is similar to the one established in Illinois in 1997.

Waste Management

Just as other states have pursued a variety of energy efficiency initiatives, most have also promoted efforts to increase their rates of solid-waste recycling. This is reflected in national rates of recycling that have climbed steadily until recent years. Few states, however, have pursued solid-waste recycling, for both household and industrial goods, with the intensity of New Jersey over the past decade and a half. In addition to familiar benefits such as reduced burden on landfills, steady increases in a state's recycling rate may also serve to contain its overall rate of greenhouse gas emissions. However, New Jersey is one of a small number of states to formally link its ongoing recycling and waste management efforts with its overall strategy to reduce greenhouse gases. This emphasis reflects the origins in solid-waste management and recycling for several DEP staff who have assumed key roles in greenhouse gas reduction.

The recycling of solid wastes has considerable potential to reduce emissions of greenhouse gases, particularly methane. Analyses by the EPA indicate that, on average, 1.67 MMT of CO_2 equivalents are avoided for every MMT of municipal solid waste that is recycled by reducing the venting of methane through landfills and the generation of carbon dioxide through incineration. Methane has a particularly powerful greenhouse impact, with a global-warming potential that is twenty-one times greater than an equivalent amount of carbon dioxide. Methane occurs naturally through the decomposition of organic material that is a major by-product of agricultural practices and solid-waste landfills. It contributes an estimated 9 percent of total New Jersey greenhouse gas releases, and 72 percent of that total comes from landfills.[28]

Those levels would be even greater if New Jersey had not already implemented aggressive solid-waste recycling programs, beginning in the 1980s and expanding in the early 1990s.[29] These rates served to increase the state recycling rate from 34 percent in 1990 to 43 percent in 1997. Consequently, New Jersey already claims credit for its post-1990 increases, and their attendant reduction of total greenhouse gases, under its Action Plan. At the same time, the plan endorses a much higher recycling rate—60 percent by 2005, and it has limited formal power to influence recycling, so new legislation and substantial financial subsidies will probably be required for the state to meet this target, given the stagnation in many solid-waste markets in the eastern United States. One major recycling emphasis linked to the Action Plan was plastics recycling, through expanded efforts to develop a standardized state protocol for this elusive component of solid-waste recycling and measurement of potential greenhouse gas impacts.

However, the plan has served to expand and intensify efforts to reduce methane releases through new management practices for existing landfills. New Jersey is indeed heavily populated with landfills, many of which have long since been closed through a process of consolidation into larger, more modern facilities. Most landfills—in New Jersey and around the nation—have never been carefully assessed for their methane releases; many simply vent these gases into the atmosphere. As Robert Shinn noted in his speech to the 1998 ECOS meeting in Wisconsin, "We traditionally thought of landfill capping in terms of protecting groundwater and surface water. From a CO_2 perspective, capping also eliminates VOCs [volatile organic compounds], methane (which has a twenty-plus greenhouse gas multiplier), and CO_2 itself. It makes perfect sense to coordinate programs that have these obvious cross benefits."[30]

Unlike other greenhouse gas elements, methane emission is amenable to control through readily available methods that can either recapture it, preventing its release and using it to generate electricity, or convert it into carbon dioxide, a less-potent greenhouse gas. New Jersey's linkage between methane and greenhouse gases has led to a significant expansion of these methane recovery and containment efforts. This began through a prioritization process whereby the DEP funded the closure and remediation of twelve large landfills with significant methane releases. Various funding sources have been tapped for this work, including excess funds from the New Jersey corporate business tax as well as other public sources such as state dredging and infrastructure trust funds.[31] "As we got through the inventory, it became obvious just what a problem methane was for us," one state official recalls. "That's what led to a major new emphasis on landfills, both closure programs and tapping methane sources."

Remediation of these twelve sites is expected to reduce the state's methane emissions by 7 percent a year. At least forty-seven additional facilities remain as candidates for such treatment in the future and are specified as significant priorities in the Action Plan. The plan acknowledges the financial hurdles to achieving comprehensive methane capture or closure of these varied facilities but outlines a number of technologies and policy tools that could accelerate this transition, particularly for landfills that may be approaching closure in coming years.[32] In fact, New Jersey projects that 22.7 percent of its total pledged reduction of greenhouse gases will come from waste management practices, through a combination of recycling and methane recovery (see table 4-1).

Leading by Example

Governor Kean's 1989 executive order calling upon state government institutions to take the lead in finding ways to reduce the greenhouse gases released by their own activities remains in force. The Action Plan has sustained that initial call but added considerable specificity to it. Perhaps the most visible effort to engage state government in this process has been active lobbying by Action Plan leaders to encourage various state departments and units to sign covenants of sustainability. In response, major state entities such as the Departments of Transportation and Community Affairs have signed such covenants and have begun to explore ways to reduce emissions caused by their facilities and programs. The Department of Community Affairs, for example, has substantial responsibility for overseeing housing policy, including the development and rehabilitation of housing for moderate- and low-income citizens. One consequence has been the amending of its rules for new construction of subsidized housing. In response to the tremendous escalation of housing costs in New Jersey in the past decade, the state has created a series of programs to subsidize construction costs for lower-income groups. Much of this funding comes from the transfer fee placed on New Jersey real estate transactions. To secure this funding, construction firms must specify how their proposals adhere to the goals of the Sustainability Greenhouse Gas Action Plan. Invariably, this has led to a review of methods for maximizing energy efficiency, and it has already served to increase emphasis on greenhouse gas reduction options in all aspects of New Jersey residential construction. In turn, the department has begun to explore other policies that might further improve residential energy efficiency.

New Jersey has also examined the ways in which state agencies consume energy in completing their daily tasks. This has included a major review of energy efficiency in all state government buildings and purchase of an extensive fleet of low-emission vehicles for use by state officials. One formal initiative includes a state policy to obtain 15 percent of the total electricity that it uses from renewable sources. This has involved the creation of a multidepartment task force known as the New Jersey Consolidated Energy Savings Program to guide the purchase process. This program has established a bid solicitation process to guide the purchase of renewables and identified funding to cover the current differential in cost between renewable and traditional energy sources. In addition, the state agreed in May 2002 to purchase approximately 12 percent of the electricity it plans to use from "green" sources.

Reaching beyond State Boundaries

New Jersey's approach to climate change has entailed a range of intrastate strategies but has also been explicit about the limitations of approaching the issue in a unilateral way. New Jersey has only 3 percent of the total population of the United States and produces only 2 percent of the nation's total greenhouse gas emissions,[33] so state officials have been unusually attentive to opportunities to build partnerships with other governments, both domestic and international. These partnerships have been intended both to maximize the potential effectiveness of New Jersey's own greenhouse gas reduction efforts and to build a larger base to establish interjurisdictional cooperation on these issues. This issue of expanded collaboration across state boundaries may constitute an important opportunity for future state policy development in this area.

Federal Government Relations

Intergovernmental funding is a well-established tool that allows state innovation, even in the area of environmental policy, where transfer dollars have been few in contrast with proliferating federal-to-state regulatory commands. But few states have proved as successful as New Jersey in tapping into available federal sources in the area of greenhouse gas reduction. Although the Whitman administration did provide significant initial funding for Action Plan implementation, the federal government has been a major funding source both for New Jersey's early activities and for some of its more recent efforts.

Federal funding on climate change began to flow to New Jersey in the mid-1990s, when the state secured grants to conduct a greenhouse gas inventory and underwrite many of the costs of Action Plan development. "Someone noticed this federal grant opportunity one day," recalls a DEP official. "We got the grant, and it really got the ball rolling, especially given all the other budget cuts facing the department." Subsequently, the state has received more-specialized funding to cover key elements of new policies related to methane quantification and recovery, expanded greenhouse gas data analysis, and exploration of possible trading strategies for carbon emissions. The DEP and the BPU have also actively explored ways to engage state-based firms in other federal programs that could reduce greenhouse gases. For example, the state secured federal grant support to assist firms participating in voluntary U.S. Department of Energy programs that provide technical assistance on energy efficiency and improved environmental performance.

New Jersey also decided to probe just how serious the Clinton administration was in its proposals to devolve additional regulatory authority to states in exchange for superior environmental performance. The DEP was an unusually early and active participant in the National Environmental Performance Partnership System. Somewhat analogous to Project XL, NEPPS was another cornerstone of the environmental policy component of the Clinton "reinventing government" strategy. It was intended to foster intensive intergovernmental bargaining, whereby states could be given considerable latitude to establish their own policy priorities in exchange for demonstrable evidence that they were achieving environmental outcomes superior to those that would occur under traditional, command-and-control approaches. The program did not allow intergovernmental negotiators to erase existing federal statutes, but it reflected a federal willingness to put many traditionally confrontational issues on the table. These included procedures for permit issuance, experimentation with pollution prevention strategies, and intrafacility flexibility in capping overall emissions rather than exact standards imposed on each and every single point source of pollution in a given facility. In turn, states that engaged in this process were expected to demonstrate a willingness to fully develop and implement their experimental proposals and formally commit to significant overall improvements.

In many respects, NEPPS was a disappointment. On the one hand, the federal government often proved reluctant—as it had in the case of Project XL—to give much ground from traditional practice. It is not clear that NEPPS had considerable support below the very highest levels in the Clinton-era EPA, and its potential impact was thereby limited. On the other hand, many states proved quite timid in proposing new strategies and unwilling to develop publishable measures of their environmental performance. Particularly effective states proved reluctant to be potentially perceived by home constituencies as overly active in comparison with others; less effective states did not want to be portrayed as laggards. Another common concern of states under NEPPS is that they perceived themselves as subject, in effect, to "double reporting," expected to sustain conventional practices while simultaneously adhering to a new, rather poorly defined, alternative order.[34]

In contrast, however, New Jersey moved aggressively into NEPPS negotiations. It was one of the first six states to pilot the NEPPS system, and it completed Performance Partnership Agreements with EPA in 1996, 1997, and 1999. These agreements included an unusually detailed set of proposals and suggested a host of possible innovations. The DEP linked a number of these to proposals to alter its use of federal funds and thereby better align funding with evolving state priorities. This possibility of fiscal reallocation was a further component of NEPPS, under the auspices of Performance Partnership Grants. A National Academy of Public Administration study that examined the NEPPS experience in a number of states notes that New Jersey entered the process unusually well prepared and was thereby "more equipped than many other states to be a leader in making NEPPS work."[35]

Perhaps the most distinguishing feature of the New Jersey NEPPS experience was that the state used the agreements to make greenhouse gas reduction an explicit goal and further frame climate change as a serious state environmental policy concern. The mere reference to these issues set it apart from virtually all other state NEPPS proposals.[36] More important, New Jersey also decided to test whether some level of greenhouse gas reduction could in fact be incorporated into its overall strategy to restructure traditional regulations. One prominent effort built on earlier state innovation in permit integration by adding greenhouse gas reductions into the process. During the late 1980s and 1990s, the DEP had experimented with a program that offered participating firms substantial increases in operational flexibility in exchange for close scrutiny of pollution prevention opportunities and significant improvements in overall environmental performance. This began as a pilot program but was later expanded under state statute to involve more than a dozen firms around the state. One of the biggest benefits to participants was the DEP's willingness to integrate disparate permits for air, land, and water regulations into a single, unified document that guided an entire facility, contrary to the traditionally fragmented nature of permitting. The facility-wide permitting program received national attention upon completion of early cases, which revealed fundamental flaws in the existing system. In these cases, firms that were in compliance with existing regulations were found to have major emission sources that had never previously been detected, numerous missed opportunities for pollution prevention and energy efficiency gains, and transfer of pollutants across single media (air, land, and water) instead of an integrated reduction.[37]

The program, however, stumbled on political conflict and high administrative costs. Whitman budget cuts played havoc with the specialized staff recruited to implement the permit program. The spouse of the program's architect and chief administrator was a leading critic of Whitman's environmental record, which led to a political backlash against her by senior Whitman advisers. Costs to implement the integrated permits exceeded estimates, reflecting the complexities involved in these initial cases. Consequently, facility-wide permitting began to be phased out, although it was partially reincarnated under the auspices of the New Jersey NEPPS proposal.

Under this plan, known as the Silver and Gold Track Program for Environmental Performance, the state sought federal approval to create three separate tracks to guide regulatory oversight. Nonparticipation implied regulatory business-as-usual, the first track. For firms that wanted to explore regulatory alternatives, the remaining tracks offered varied benefits in exchange for demonstrated environmental improvements. Participation in both the silver and gold tracks included requirements that participants would achieve the Action Plan reduction goals, 3.5 percent below 1990 emission levels by 2005. In effect, a firm would have to sign a covenant and meet those targets as one of the conditions for gaining silver- or gold-track status. The inclusion of greenhouse gases in such a process reflected both the state's concerns about climate change and its perception that greenhouse gases such as carbon dioxide can serve as "umbrella pollutants," reduction strategies for which spill over to reductions in other emissions such as air toxics. Respective tracks also involved other provisions, including pledged reductions beyond current standards of conventional pollutants. In exchange, participation in these programs offered varying degrees of flexibility in regulatory compliance, including the possibility of securing faster and longer-term permits.

New Jersey secured EPA approval to go forward with this system and began negotiations with a number of firms that sought either silver- or gold-track designation. Ten firms gained silver-track status in 2000 and 2001. "There is enormous potential here, but there was all kinds of jockeying back and forth on what should—and should not—go into this kind of a program," a DEP staff member explains. "But adding the greenhouse gas piece is very interesting. It really goes beyond the traditional areas that are negotiated. You get to these wonderful moments where a company may say, 'We can't cut our energy use.' Then, suddenly, they see how they can, making money and reaching the covenant target in the process."

Under both the track system and the earlier experiment in facility-wide permitting, New Jersey found that it could begin to move beyond existing federal standards through a formal negotiation process. This demonstrates the possibilities of weaving greenhouse gas reduction commitments into more traditional regulatory arrangements, with support from both federal and state parties. However, the track system proved difficult to formalize within the state, and the DEP decided to eliminate the gold track in 2002 owing to complexities in interpretation and implementation. The future use of this approach, or some variation, under the new administration headed by Governor James McGreevey and DEP commissioner Bradley Campbell, remains uncertain.

International Relations

New Jersey has also ventured beyond American shores in search of intergovernmental partnerships related to greenhouse gas reduction. The long-standing relationship between the state and the Netherlands proved a logical area of exploration. Their earlier emphasis on conventional pollutants was expanded to include greenhouse gases through a series of bilateral meetings during the 1990s. These discussions culminated in a formal "letter of intent" signed in Trenton on June 5, 1998, between Commissioner Shinn and his Dutch counterpart, Margaretha deBoer, the Netherlands' minister of housing, spatial planning, and the environment. "This got a little tricky, and the State Department was very nervous about us working so closely with another country," explains a DEP official. "Ultimately, they would not let the governor sign but didn't block the agreement."

The letter of intent formally established plans for a cooperative relationship between the two governments on greenhouse gas reduction. It reflected the common interest of the signatories "to collect practical experiences in several ways to gain better insights into the Kyoto Protocol such as emissions trading, joint implementation, and the use of policies and measures" to reduce greenhouse gases. In particular, the letter of intent outlined plans to "identify trading mechanisms and identify and design a prototype greenhouse gas emissions trade" and to "collaborate on the design and implementation of an emission banking system."[38]

Both parties were particularly eager to experiment with emissions trading. They hoped to develop a series of pilot projects to gain experience and potentially move toward an active trading system. Just as New Jersey had borrowed the covenant system from the Netherlands, in this instance the latter was clearly eager to tap into American experience in emissions trading that had been developed for such pollutants as sulfur dioxide. "We know New Jersey has a very progressive policy, and in the Netherlands, we can learn from them," explains Pieter Verkert, the environmental consul at the Dutch embassy in Washington. "The agreement is to find out how the credit mechanisms will work in the future."[39] New Jersey planned to use its Open Market Emission Trading (OMET) system, which had been established in 1996, to list reductions of conventional pollutants. But more than a registry for potential future credit, OMET was intended to allow trading of the reductions in the form of credits to other parties that might use them to comply with existing regulations.

The Open Market Emission Trading system was expanded in 2000 to also allow for greenhouse gas emission reductions. The first official Notice of Generation of such credits was submitted by Onsite Sycom in January 2001. Discussions began about possible New Jersey–Netherlands greenhouse gas emissions trades, including a possible trade of methane credits between a pair of private landfills that operated on opposite sides of the Atlantic Ocean. Under another proposal, the Netherlands intended to cover the costs for a coal-fired power plant in Jersey City to convert to natural gas. This switch would reduce the plant's carbon dioxide emissions by approximately one-third and allow the Netherlands to obtain credit for emission reductions.

Much like the tracking system for permits, however, this proposal has had a rocky history. The Environmental Protection Agency funded DEP development of greenhouse gas quantification protocols but never officially approved OMET. In turn, environmental groups regularly offered concerns about the loose procedures for listing and trading credits. These criticisms were focused more on conventional pollutants than on greenhouse gases, but they raised doubts about the viability of the entire system. In turn, the Dutch began to lose interest in attempting pilot projects as it realized the uncertainty of OMET's status and as American involvement in the Kyoto Protocol proved increasingly unlikely. "They knew our 2005 goals, and that really made trading attractive to them when we signed the letter in '98," Shinn recalls. "We were working to make our system compatible with Europe. But the U.S. attitude toward Kyoto was pretty damning for them to continue."[40] Finally, the McGreevey administration announced plans to close OMET in September 2002. Commissioner Campbell has noted a history of questions about accuracy in monitoring OMET credits and declared the program "an experiment that failed."[41]

Future Definition of Prime-Time Status

The 2002 decisions by the McGreevey administration to eliminate the gold track and OMET fostered uncertainty about the next generation of climate change policy in New Jersey. These actions raise questions about the status of other elements of the state's Action Plan and about the state's future prime-time status: Will the new administration further develop New Jersey as a prime-time player with new policies, or will it begin to backtrack on the state's commitment to climate change? "There is certainly no backtracking in New Jersey, but the state hasn't been that visible on the issue since McGreevey took office," notes one regional EPA official. "They clearly want to recapture their preeminence on climate change."

Many of the provisions of the state's climate change strategy have only begun to be implemented in the past year or two, but New Jersey does appear to be on target to meet its 2005 goal. This is most likely attributable to a combination of factors, including the early impact of the emission reduction programs, the recent decline in economic activity, and increased reliance on nuclear power in the total mix of electricity generated in New Jersey. As existing programs mature, officials project that New Jersey may actually exceed its 2005 commitments, a prediction that has fueled speculation that the state may seek an even higher reduction goal or extend reduction commitments beyond 2005.

The only new initiative on climate change to emerge since the respective departures of Whitman in 2001 and Shinn in 2002 has been the creation in early 2003 of a mandatory program for reporting carbon dioxide emissions. This program is similar to the one created in Wisconsin in 1993, involving an amendment of existing air emission reports. It was endorsed in the Action Plan and developed considerable momentum in 2000 and 2001, when it underwent a formal process for review and public comment. Department of Environmental Protection officials believe that it will provide a more detailed data system that could be utilized both in tracking compliance with covenant pledges and in guiding new initiatives.

The possible existence of a new data source, along with the existing infrastructure of climate change initiatives, poses interesting challenges for the McGreevey administration. The new governor inherits an active set of greenhouse gas reduction initiatives and a large set of policy entrepreneurs in the DEP and the BPU who are eager to move ahead. At the same time, the administration faces a massive fiscal deficit and clearly wants to put its own stamp on state environmental policy. Indeed, New Jersey has a strong tradition of new administrations' attempting to aggressively discredit their predecessors' environmental records and proposing something markedly different.[42] "Greenhouse gas reductions in this state have been so identified with Shinn that there has been concern the new administration will come in and throw it out," notes the head of one prominent state environmental group. In turn, a representative of one of the private institutions that signed a covenant notes, "We don't know what to expect from the DEP. There has been this side that is trying to be more flexible, but the department is still divided into a lot of cells that don't talk to each other or trust each other. I think that's one of the reasons you haven't seen more companies sign the covenant or sign up for the tracks. We never know what is coming next."

To date, the Campbell-led DEP has been most visible in its criticism of the Shinn and Whitman era in areas outside the domain of climate change. In particular, Campbell has alleged a pattern of soft enforcement of existing legislation, particularly in areas such as wetlands and air-quality regulation, and has promised a more aggressive form of environmental management. He has also strongly criticized Bush administration proposals on climate change and air-quality regulation, noting that the Bush Clear Skies initiative "might more aptly be called Brown Skies, particularly in New Jersey."[43]

At the same time, he has indicated strong support for building on earlier initiatives, including facility-wide planning and efforts to reduce medium-specific barriers, that could be included in the next generation of state greenhouse gas reduction actions.[44] He is clearly attentive to the possibilities of building on the existing foundation of state efforts. "The actions of the Bush administration on Kyoto give states like New Jersey some political space to try to build the infrastructure for greenhouse gases," he explains.[45]

Two distinct options have emerged as New Jersey officials consider possible future directions. Although these options are not necessarily mutually exclusive, they outline different approaches that the state might take in building on the foundation established during the past several years. They involve either collaboration with or emulation of recent strategies developed by states outside the pool of cases examined in this study. These new initiatives are in the earliest stages, but both have received considerable attention in the state as models for possible next steps.

Going Regional

New Jersey has a long history of interacting with neighboring states on environmental concerns. In some cases, these relationships have been highly acrimonious. New Jersey and New York have battled in court for more than seven decades over solid waste, usually reflecting New Jersey's contention that poor waste management practices in New York result in considerable pollution of its coasts by garbage and medical wastes. In turn, states such as New York and Connecticut have long alleged that New Jersey has provided tacit support for air and water pollution "export," proving far less attentive to regulation in cases in which pollutants are likely to cross state boundaries.[46] New Jersey has also reached common cause with neighboring states on other issues, however, such as instances in which midwestern air emissions ultimately posed serious problems for New Jersey and its northeastern counterparts.

New Jersey now has the option of joining forces with its New England neighbors in expanding the physical zone of its efforts to reduce greenhouse gases. During the summers of 2001 and 2002, the governors of the six New England states and five eastern Canadian provinces reached formal agreement on three goals for reduction of regional greenhouse gas emissions. These begin with a goal to reduce regional emissions to 1990 levels by 2010, essentially pledging these eleven jurisdictions to honor the Rio Declaration targets. This is followed with a goal to reduce regional emissions "by at least" 10 percent below 1990 levels by 2020. This provision also established "an interactive five-year process, commencing in 2005, to adjust the goals if necessary and set future emission reduction goals."[47] Finally, they pledged to reduce regional greenhouse gas emissions "sufficiently to eliminate any dangerous threat to the climate; current science suggests that this will require reductions of 75–85% below current levels."[48]

This intergovernmental and international agreement reflects nearly three decades of formal collaboration between these respective governments, including previous environmental accords on reducing such conventional pollutants as sulfur dioxide and mercury. The chief elected officers of these governments meet annually and have produced an array of common policy initiatives during this period. However, the issue of climate change and greenhouse gas reduction was a relative newcomer to these sessions, first introduced in 2000 during regular meetings that took place between officials from the environmental and energy departments of the various provinces and states. The very suggestion for a common regional target appears to have emanated from an energy ministry official in New Brunswick, and the proposal won broad support from colleagues from other agencies and ministries. This followed the pattern of agency-based entrepreneurship evident in other cases, demonstrating the possibilities of extending this process of policy development across single-state boundaries. The energy and environmental officials prepared the resolution for the governors and premiers, who gave it their unanimous support in August 2001.

This approval launched a process in which state and provincial officials intensified their interactions and began to map out some common goals that might facilitate attainment of the reduction goals. The agreement recognizes that the goals are established for the overall region and that they "may not be achieved in equal measure by each jurisdiction." Nonetheless, the jurisdictions have begun to develop regionwide strategies for energy efficiency standards, common purchase and use of highly efficient technologies by state, provincial, and local governments, and a detailed inventory of all regional emissions. In time, they would clearly like to establish a system for cross-boundary emission trading. An initial set of pilot projects and affirmation for continued policy development were endorsed by the governors and premiers during their annual meeting in Quebec in August 2002.

Officials from all of the states and provinces involved in this agreement are adamant that they want this partnership to go forward even as Canada begins to prepare to honor its November 2002 ratification of the Kyoto Protocol. But they are also clear that they would like to expand the zone represented in the agreement to build a larger base for policy development and implementation. Collectively, the New England states of Connecticut, Maine, Massachusetts, New Hampshire, Rhode Island, and Vermont produce approximately 3 percent of total U.S. emissions. Their Canadian provincial partners, New Brunswick, Newfoundland, Nova Scotia, Prince Edward Island, and Quebec, generate approximately 15 percent of total Canadian emissions. American and Canadian greenhouse gas emissions are roughly comparable on a per capita basis, so the existing agreement represents approximately 5 percent of total emissions from the two nations.

Consequently, the idea of expanding the boundaries of this compact is highly desirable from the standpoint of the eleven jurisdictions. "We have had probes of interest from some states and have previously worked with states like New Jersey and New York on other environmental issues," explains a senior official of the Conference of New England Governors. "Under our governance structure, we could easily sign on additional states as adjunct members and formally include them in everything that we do."

In turn, the secretariat that represents the five eastern provinces also views future expansion with enthusiasm. "For the most part, these are small states and provinces, so we clearly gain more expertise and leverage by working together," a Canadian official has stated. "We need this type of relationship and realize these energy and environmental issues transcend state and provincial, and even national, borders."

New Jersey is clearly a desirable negotiating partner, given its early and active engagement on climate change issues. New York, Maryland, and Pennsylvania have also been increasingly active in developing their own greenhouse gas reduction programs. As a result, leaders of the New England–eastern Canadian pact are eager to explore ways in which their agreement might evolve into a physical zone that represents a significant portion of North America and allows for creative policy development across boundaries. Although New Jersey's international environmental engagement has been most intensely concentrated on the Netherlands, it has also been active historically in deliberations with the federal government of Canada. In fact, during the past decade the state entered into formal agreements with Canada that "promote the exchange of environmental technologies and information." These involve formal Memorandums of Understanding whereby both governments agree to expedite reviews of technologies for various environmental protection or energy efficiency programs for mutual use. As former commissioner Shinn and DEP official Matt Polsky have written, the state has established "a good working relationship with the Canadian government and some businesses and, through our agreement with Canada, we are developing a prototype for a technology verification system."[49]

Consequently, New Jersey could indeed expand the scope of its efforts by bringing its considerable expertise on climate change policy into this regional agreement. Moreover, state environmental and energy officials have long sought a more formal working relationship, much as exists for the eight-state Northeast States for Coordinated Air Use Management, which addresses regional air-quality issues. The organization "already includes New Jersey and New York, so this provides a perfect opportunity for us to bring them into our group," noted a Canadian provincial official. The Northeast States for Coordinated Air Use Management is now headed by Kenneth Colburn, one of the key entrepreneurs behind New Hampshire's climate change initiatives and an early advocate of the New England–eastern Canadian accord. Even the regional office of EPA is likely to be a strong supporter, because it is now directed by Robert Varney, who headed the New Hampshire Department of Environmental Services during that state's active movement into climate change policy. Moreover, the New England regional office of EPA has been heralded as "one of the most dynamic centers of innovation" in the agency.[50] Were New Jersey to join forces with the other participating jurisdictions as well as New York, Maryland, and Pennsylvania, this expanded zone would cover approximately 15 percent of total American and Canadian emissions of greenhouse gases. It would clearly exceed the emissions of the largest subnational emission source in North American, Texas (see table 1-1).

Going Regulatory

New Jersey officials have also begun to wrestle with the question of whether future steps to reduce greenhouse gases need to involve more traditional tools of regulation. Much of the existing New Jersey effort is based on collaboration and voluntary commitments, such as the covenant system, although many of its key energy provisions have a statutory—and regulatory—foundation. The state could clearly build on these elements of its strategy. For example, McGreevey has already proposed that the state double its regulatory standard for renewable energy, reaching a level of 13 percent by 2012. At the same time, additional measures are clearly under consideration in Trenton. "We may be getting to the point where things need to be a little more regulatory," explains one DEP official who has been active in all stages of Action Plan development.

Whether or not it acts in concert with other neighboring states, New Jersey has a series of options should it take a more regulatory route. Indeed, the McGreevey administration appears more philosophically inclined toward regulatory as opposed to voluntary approaches than was its predecessor, a preference reflected in the decisions to terminate OMET and the gold track as well as much of its early rhetoric. In his former role as a regional administrator of EPA, Campbell played a prominent part in blocking Wisconsin's negotiated agreement with a major utility that included greenhouse gas reductions, citing conflict with federal statute. Consequently, regulation-based approaches, such as multipollutant strategies that would concentrate on electrical utilities and follow the examples of New Hampshire and Massachusetts, may be preferred in the administration.

One option that has begun to receive attention in the McGreevey administration involves new land-use policies that could reduce sprawl and reliance on long-distance commuting. This represents a move beyond Whitman-era initiatives, most notably a $1 billion bond issue approved in 1998 and designed to purchase a million acres of rural land over a decade and close the land to development. Under McGreevey, the state has launched a program entitled Blueprint for Intelligent Growth. Consistent with Maryland and a few other states that have developed ambitious "smart growth" strategies, the program is intended to channel future growth into already-developed areas, including central cities and established suburbs. It divides all of New Jersey into three categories, endorsing growth in areas adorned with the color green on its statewide map. This program is not expressly designed to reduce greenhouse gases but could indeed, if implemented, have such an impact. It has, however, already proved controversial and has led to a number of early adjustments before any implementation effort has been made.[51]

The state may also be exploring ways to develop regulation to address transportation-generated emissions of greenhouse gases. Transportation currently generates approximately 30 percent of New Jersey's annual greenhouse gas emissions, yet it has proved to be the sector most difficult to address in any meaningful way. "We feel that we are really rolling in some of the other key areas, especially energy, but trying to craft a strategy for transportation is very hard," admits one senior state official. As in other states, New Jersey has found relatively little support within the Department of Transportation to take greenhouse gas reduction seriously. This appears to reflect the traditional emphasis in such departments of promoting more driving through highway expansion. "Transportation was the most frustrating area of my eight years in trying to get something started related to greenhouse gases," laments Shinn.[52]

Of course, the very idea of regulatory options concerning transportation has assumed new definition since July 2002, when the state of California decided to attempt to regulate the carbon dioxide emissions of motor vehicles. There are numerous legal questions of state government jurisdiction over such matters, but New Jersey and other northeastern states are clearly beginning to consider legislative emulation of the California strategy. For example, New York Republican governor George Pataki noted in his 2003 State of the State address, "Let's work to reduce greenhouse gases by adopting the carbon dioxide emission standards for motor vehicles which were recently proposed by the State of California."[53]

California reassumed its historic role as a national leader in air emissions when AB 1493 was signed into law by Democratic Governor Gray

Davis on July 22, 2002, in dual ceremonies in Los Angeles and San Francisco. The legislation launches a process whereby the California Air Resources Board (CARB) will develop regulations by January 1, 2005, that will achieve "the maximum feasible and cost-effective reduction" of greenhouse gases from cars, sport-utility vehicles, and light-duty trucks. These regulations would go into effect one year later, after a period in which the legislature could review and amend them. Vehicle manufacturers would be expected to implement changes for their 2009 models, although they could secure credits for any emission reductions achieved between 2000 and the start of the 2009 model year. The California Air Resources Board is an eleven-member body appointed by the governor.

This new policy, however, remains very limited in details, particularly in comparison with the kinds of policies that have been developed in other states. The legislation does not establish any specific emission reduction targets or goals and gives no guidance to CARB on how reductions might be achieved. It also prohibits CARB from imposing any new taxes, reducing speed limits, banning categories of vehicles, requiring any reduction in vehicle weight, or imposing mandatory trip-reduction plans on motorists in developing its regulatory strategy. Instead, CARB is expected to review a host of other approaches and develop a comprehensive plan by 2005.

California's actions are based on its long-standing autonomy in setting air-quality standards. This was made possible by CARB's creation in the 1960s, which preceded the formation of EPA and built on a series of earlier state laws intended to reduce air pollution. The state has historically used this "unique loophole" to take action independent of the federal government, which may then be emulated by other states.[54] California led the nation in earlier initiatives regarding the use of unleaded gasoline, catalytic converters, and clean diesel fuel, establishing a precedent for the greenhouse gas reduction legislation. In these instances, the action by a state that registers approximately 10 percent of the nation's noncommercial vehicles ultimately led to federal action to establish a more uniform standard for the nation. "As California goes, at least in air pollution, so goes the nation," the political scientist Donald F. Kettl has noted.[55]

The new vehicle legislation also builds on a series of earlier California initiatives that were either explicitly or implicitly intended to reduce greenhouse gases. These include an extensive series of programs to promote energy efficiency and renewable energy, comparable to those of other states introduced in this and earlier chapters but unusual in their large number and overall range. Many of these efforts have had strong bipartisan support and have, collectively, helped the state secure one of the lowest rates of per capita release of greenhouse gases in the nation (see table 1-1). California also enacted legislation in 1990 that required that 10 percent of all cars sold in the state by 2003 have either literally or virtually no emissions, most likely encouraging the development of electric cars. Implementation, however, has been dogged by delays in developing the needed technology and a series of legal challenges, leaving California unable to even approach that goal by the original deadline. In addition, the Bush administration's Justice Department joined vehicle makers in challenging the constitutionality of this legislation in October 2002, contending that it usurped federal authority for setting fuel efficiency standards in vehicles. Auto manufacturers contend that the standard has forced them to invest heavily in technologies with little short-term application, thereby reducing their ability to focus on more viable alternatives that further reduce but do not eliminate emissions.

This legislation is also significant in that it reflects a very different kind of political process from most of the greenhouse gas reduction programs introduced earlier in this study. Whereas most of these other efforts involved a relatively quiet process of policy development and consensus building, nurtured by state agency officials, the California case reflects a more traditional and adversarial style of environmental policymaking. It represents a familiar divide between environmental groups and auto manufacturers "in a clash reminiscent of their first collisions more than three decades ago."[56] The California legislation reached Governor Gray Davis's desk only after intense, highly partisan debate in both chambers of the California legislature, resulting in a one-vote margin of victory in the Assembly that closely followed party lines. "This bill represents the worst form of environmental extremism," said Republican Assembly leader David Cox, of Fair Oaks, in leading the opposition.[57] Davis did not signal his views on the bill or announce his willingness to sign until after its passage.[58] The bill was drafted by an environmental group, Bluewater Network, and sponsored by Democratic assemblywoman Fran Pavley, a first-term legislator and former civics teacher from Agoura Hills. It was backed by a coalition of environmental groups, Silicon Valley business leaders, and large California municipalities. It also received strong support from prominent entertainers, who were active at the signing ceremonies, and former president Bill Clinton.

The implementation of this legislation, however, remains highly uncertain, owing to its vagueness, the potential for a future legislative veto of

CARB actions, and strong interest group opposition. A multimillion-dollar lobbying campaign failed to block the legislation, but a coalition of major industries has launched a legal challenge alleging that the bill constitutes an excessive extension of CARB authority. The Bush Justice Department may join this suit, as it did in October 2002 on the earlier legislation for low-emission vehicles.[59] Opponents initially weighed the possibility of a November 2002 ballot initiative to repeal AB 1493, consistent with the state's extensive use of direct democracy tools to resolve previous environmental and energy controversies. They ultimately decided against this for 2002, given short deadlines to secure more than 420,000 petition signatures and the decision to lead with the court challenge. Opponents vowed repeal of the legislation in campaigning leading to November 2002 legislative and gubernatorial elections, although Davis won reelection and Democrats retained control of both chambers of the legislature. The repeal effort has since been overshadowed by California's gargantuan fiscal crisis and the ultimately successful attempt to recall Davis that focused primarily on his handling of the fiscal situation and the earlier shortages of electricity.

Clearly, the California legislation offers an alternative model for New Jersey, as well as other prime-time states. It is uncertain how such a legislative proposal would play before the state legislature in Trenton, but a number of northeastern states have clearly begun to consider this as a model for a more regulatory approach. However, transportation may be an even more sensitive issue for the McGreevey administration than for its predecessor, given a 2002 decision by the Ford Motor Company to close an auto manufacturing plant in the Middlesex County town of Woodbridge, which McGreevey formerly served as mayor. Nonetheless, transportation remains one gaping hole in New Jersey's previous efforts to curb greenhouse gases, and the state is uniquely positioned to help define the next generation of state policies toward climate change.

Lessons from New Jersey

Like other prime-time states, New Jersey has been explicit in framing climate change as a serious environmental threat that warrants a multifaceted response. This has enabled the state to be explicit in labeling its intent to reduce greenhouse gases through a number of initiatives. Much like its New England and California counterparts, New Jersey has gone considerably further in policy development than most other states and perhaps many other developed nations. It also demonstrates the potential for policy development when a strong coalition forms between agency staff and its director, who then work to secure support from elected officials. It further demonstrates the possibility of adopting ideas commonly employed on the other side of the Atlantic for use in the United States, given its reliance on the covenant system developed in the Netherlands and celebrated in much environmental analysis.[60]

At the same time, the New Jersey case raises as many questions as it answers. Can its emphasis on collaboration and voluntary strategies deliver greenhouse gas reductions over the long term? Are its programs sustainable, given the historic tendencies in the state toward "regime change" that tends to be very disruptive of the policies of previous governors? Can this approach be linked with those of other states, such as potential New England partners? Will it evolve toward the more divisive and adversarial pattern seen in the California case?

CHAPTER 5

Looking Ahead to
the Next Generation
of Climate Change Policy

Each session of Congress brings new expectations that political conflict may yield long enough to allow enactment of some significant federal effort to reduce greenhouse gas emissions. Such a step might begin to establish a long-term policy architecture for addressing the threat of climate change after more than half a decade of fruitless debate over possible American engagement in the Kyoto Protocol. Prospects brightened for a time during the 107th Congress, as the Senate passed a national version of a renewables portfolio standard already in operation in Texas, Wisconsin, New Jersey, and a dozen other states. Although it was loaded with loopholes and exception clauses, this provision endorsed a national standard that would require that 10 percent of American electricity come from renewable sources by 2020. It appeared headed for a serious hearing in conference deliberations over a gigantic energy bill with numerous provisions designed to increase energy supply, including drilling for oil in the Arctic National Wildlife Refuge in Alaska. Finally, it appeared, the United States was on the verge of assembling at least one piece of a greenhouse gas reduction program.

But much like its legislative predecessors, serious federal efforts to create policy to reduce greenhouse gases are a bit like the proverbial story of Charlie Brown attempting to kick the football held by Lucy. Supporters occasionally get close, but other factors conspire to prevent contact. In this instance, most elements of the energy legislation collapsed in partisan and interbranch warfare, leaving only skeletal elements of the original bill under consideration. Even President Bush's climate change fig leaf, intended to elevate the largely meaningless concept of reducing the carbon intensity of the American economy as a climate change policy, failed to receive serious consideration in 2002 and 2003. For more than a decade, this has been the story of federal efforts to reduce greenhouse gases, seemingly regardless of which individual occupies the White House or which party dominates Congress.

Since federal legislation in 1990 and 1992 established the precedent of emissions trading and created opportunities to expand development of renewable energy, the federal debate over greenhouse gas reduction has largely been characterized by a blend of hyperbole and inertia. Aside from programs targeted on financial and technical assistance for select industries to try to develop more climate-friendly technologies and sustain research in the basic sciences related to climate change, the federal government has proved largely incapable of action on greenhouse gas reduction. By 2000, presidential candidate Al Gore began to lament to campaign advisers that his earlier advocacy for action on climate change was problematic for him politically and that he needed be very cautious in dealing with it. Even had Gore won the divisive 2000 election, it is almost inconceivable that the Senate would have ratified the Kyoto Protocol and highly uncertain that Congress would have taken any serious steps to address climate change.

For other industrialized nations, this inaction has created considerable opportunity to heap opprobrium on the United States, and the Bush administration more specifically, for its seeming indifference to greenhouse gas reduction. In turn, American disengagement from the Kyoto deliberations has been linked with other shifts in international affairs to render an increasingly popular portrayal of the United States as an arrogant Lone Ranger in world affairs, using its status as the sole remaining superpower to come and go as it pleases across a range of international issues. Such a critique raises legitimate questions of future American engagement in international affairs. It is interwoven with a nagging doubt about the capacity of America's national political institutions to function and generate effective new policies in a range of domestic issue areas, from medical care to homeland security.

The performance—or lack thereof—of the American federal government has tended to dominate most analyses of climate change policy by scholars and journalists, whether they operate domestically or internationally. This is highly understandable, given the large American contribution to global levels of greenhouse gas releases and the active involvement of the Clinton-Gore administration in Kyoto negotiations in 1997. But this preoccupation has served to obscure at least two important developments that complicate the increasingly popular depiction of an international community determined to move ahead with greenhouse gas reductions in the face of American intransigence: the difficulty that Kyoto Protocol signatories have encountered in moving beyond their pledges to establish emission reduction policies and the relative success of some American states in enacting such policies.

International Outrage versus Domestic Commitment

Closer scrutiny of the nations that have ratified the Kyoto Protocol suggests that many are now facing considerable political struggles in attempting to achieve their pledged reductions. The reconciliation of international pledges with domestic politics is never easy, but the process appears particularly contentious in coming to grips with both the reduction levels set forth in the protocol and the creation of a credible long-term process to guide policy. As the political scientist Robert Putnam has long since noted, national political officials simultaneously attempt to "maximize their own ability to satisfy domestic pressures while minimizing the adverse consequences of foreign developments."[1] Given American disengagement on climate change policy, other developed nations, most notably those of the European Union (EU), have taken center stage. Although they are unified in condemning recent American actions, they have also had to confront the difficult issues of securing reduction pledges from other nations and beginning to reduce their own emissions. This has resulted in repeated softening of the terms of engagement to secure participation from individual nations, far beyond anything envisioned in Kyoto. Since its decision late in 2002 to ratify the Kyoto Protocol, for example, Canada has continued to insist that generous treatment of "clean energy credits" for exporting its natural gas to a Kyoto nonparticipant, the United States, would be pivotal to its plans to honor Kyoto reduction targets. This repeated national pattern of jockeying after ratification has led to growing concern that "the relatively stringent emissions targets negotiated in Kyoto have been so diluted" in more recent rounds that there may be a substantial delay before any significant reductions can be achieved.[2] It is also a reminder that long-term reduction of greenhouse gases will require development of multilevel governance strategies that cut across the many spheres of policy that contribute carbon dioxide and related emissions. Any international deal will be only the beginning of a long-term process that is daunting in its potential complexities of collective action.

In turn, EU nations and other signatories have had to move beyond their broad pledges. They have begun to look internally at their political commitment—and institutional capacity—to achieve pledged reductions. Recent evidence suggests that they are finding this transition difficult, given the need to "satisfy domestic pressures," using Putnam's formulation. Initially, it had been assumed that a substantial amount of the European reductions would be achieved through increased carbon taxes on various fossil fuels. As the political scientist Neil Carter has noted, by 2001 eight EU nations had established some form of carbon tax related to their planned greenhouse gas reductions. However, as the EU and some other participating nations have begun to get serious about Kyoto ratification, some member nations have begun to backtrack on their new tax instruments. Finland has already repealed its carbon tax, Sweden weakened its tax after one year, and Germany's Christian Democratic Party pledged elimination of the new German tax before its narrow defeat in the September 2002 elections. Britain and France have proved particularly resistant toward any EU-wide carbon taxes.[3] This is consistent with a pattern detected in Europe several years ago by the policy analysts Ute Collier and Ragnar Lofstedt: "At the national level, much effort has been put into dressing up measures and developments as 'climate change measures,' when in reality emission targets are likely to be missed almost everywhere."[4]

Relatedly, the European Environment Agency has reported that at least ten of the fifteen European nations that have now ratified Kyoto are likely to fail to meet their pledged share of emission reductions by significant margins. Struggling nations include Austria, Belgium, Denmark, Finland, Greece, Ireland, Italy, Portugal, and Spain.[5] Moreover, the European nation that is most commonly commended to Americans as a model for innovation on climate change and other environmental issues, the Netherlands, also falls into this camp. Indeed, the nations that remain most steadfast behind their Kyoto commitments tend to be those that anticipate the easiest transition processes. In the United Kingdom, for example, Kyoto compliance will be easy. This is almost entirely attributable, however, to a massive shift in energy policy from coal to natural gas that was launched more than a decade ago. This shift was part of a major effort to privatize electricity generation and distribution in the United Kingdom and was not linked to planned greenhouse gas emissions. Instead, it reflected a decision to move away from prior national commitments to mine and burn British coal and launch a "dash for gas" that took advantage of discoveries of substantial offshore deposits of natural gas.[6] Subsequently, British greenhouse gas emissions have declined during the past decade, largely attributable to this massive shift toward a less carbon-intensive fuel.

Intensified commitment to nuclear power as a core energy source is a central element in France's long-term strategy to reduce greenhouse gas emissions. Of course, nuclear power generates no greenhouse gases but remains highly controversial in most other Western nations because of other serious environmental concerns that it poses. In Germany, Kyoto compliance will be eased substantially by the economic collapse in the 1990s of the former East Germany after its incorporation into a unified German nation. This dramatic decline in economic output from industries with exceedingly low levels of energy efficiency has significantly closed the gap of additional reductions that Germany would be required to make.

By 2003, EU nations were clearly accelerating the pace of attempting to develop a system for securing and trading carbon credits, assembling key elements of a viable trading system. But their limited prior experience with emissions trading and reliance on a highly uncertain system of compliance measures and monitoring has placed in serious doubt just how prepared various Kyoto signatories are to ensure the meaningful reductions to which their national governments are pledged.[7]

In October 2002 the EU did propose major new funding initiatives to develop highly fuel-efficient vehicles, reminiscent of the earlier Clinton-Gore strategy to help underwrite a new system of transportation through the Partnership for a New Generation of Vehicles. At present, both initiatives remain possible long-term strategies rather than tangible sources of short-term and mid-term emission reductions.

Emergence of the States

At the same time that Europe and other potential Kyoto participants have struggled with policy design and implementation, the United States has not proved as moribund on greenhouse gas reduction as conventional wisdom would indicate. Chapters 2 through 4 of this study reveal an array of policies that have been enacted at the state level with either the express or the indirect intent of reducing greenhouse gases. These cases, though they do not constitute a comprehensive overview of the fifty states, indicate considerable potential within the American political framework to develop a wide range of policies that respond to the challenge of climate change.

Collectively, these cases present an alternative policy architecture for greenhouse gas reduction that could indeed be expanded to other states, the entire nation, or even other countries, in coming years. Rather than a singular system imposed nationally or internationally, states are beginning to test what does and does not work in reducing greenhouse gases. Many of these state programs remain in early stages of implementation, reflecting their relatively recent creation, but reduction impacts are beginning to be seen in some of them. For example, renewables portfolio standards have clearly triggered a dramatic increase in the use of renewable energy in Texas, and they are now being implemented in fifteen other states (see table 2-1), collectively responsible for more than 40 percent of total national greenhouse gas emissions. At the same time, multipollutant strategies that include carbon dioxide are being woven into new air-quality strategies in multiple states, an approach that appears likely to spread to other states in future years. Still other states are attempting such initiatives as methane recapture from landfills, sequestration of carbon from agricultural and forestry reforms, detailed reporting of carbon dioxide releases to guide next stages in policy development, and mini carbon taxes through social-benefit charges to fund energy efficiency and renewable-energy projects.

Perhaps most surprisingly, these states have generally been able to enact these new policies without the kind of divisiveness that either precludes most federal policies or clouds their implementation with likely legal and political battles. Until the 2002 legislation that set California on a path to establish carbon dioxide emission standards for vehicles, the state experience in developing climate change initiatives has generally been bipartisan and consensual. This has reflected a process of careful policy development by policy entrepreneurs, often officials employed in state government agencies. These entrepreneurs have tailored policies to the political and economic realities of their particular setting and have built coalitions that seem almost unthinkable when weighed against the past decade of federal-level experience. They have helped make possible a series of diverse policy experiments that can now be examined carefully as they move into more advanced stages of policy implementation.

In many respects, this pattern is in keeping with the traditions of American federalism. Even before the most recent era of state resurgence, states have long been incubators of policy ideas that ultimately swept across regions and, in some instances, would be embraced in some later form at the federal level. That pattern, of course, has only intensified in recent decades as state policymaking capacity has risen steadily alongside an attendant decline at the federal level. That shift raises a more fundamental question of whether states can be expected to play a more central role in long-term policy development and implementation for greenhouse gas reductions or are instead best viewed as a testing ground for early experiments that will inform future federal policy.

Regardless of long-term direction of policy, states with active climate change programs may also be contributing to a basic understanding—with potential transcontinental lessons—of how best to reduce greenhouse gases. Ironically, at the very time that the Kyoto Protocol negotiations reached their climax in late 1997, the economist Thomas Schelling published an article in *Foreign Affairs* on the challenge of climate change. Schelling notes the enormous economic and political impediments that would face any attempt to construct a comprehensive, long-term strategy for a problem that is only beginning to be understood. But he does not call for inaction. "In the short run, there will almost certainly be innumerable modest but worthwhile opportunities for reducing carbon emissions," he explains. "A program of short-term reductions would help governments learn more about emissions and how much they can be reduced by different measures."[8]

Since the publication of that article, enormous energy—political, economic, and intellectual—has been concentrated on an effort to construct the very type of comprehensive structure that Schelling questions. Far more quietly, a growing number of American states appear to have heeded his advice. Many have now assumed a role of international leadership in achieving initial reductions and providing an array of policy tools that might be refined or expanded—whether on a subnational, national, continental, or international scale—in coming years and decades. Some serious challenges face expanded development of these tools, however, in terms of their potential long-term role in defining climate change policy.

Obstacles to Further Decentralization of Climate Change Policy

Decentralization of governmental functions often proves more appealing when presented in abstract form than as a specific policy proposal. For decades, think-tank studies and blue-ribbon commissions have extolled the virtues of decentralization—in particular, the idea of devolving select federal responsibilities to state governments. These proposals have been advanced in virtually every sphere of public policy, including the environment. They tend to founder, however, in the presence of some admixture of federal reluctance to relinquish authority and state hesitation to expand duties in the absence of major new funding sources.

Nevertheless, during the past decade devolution has been pursued aggressively in some spheres of policy that had been dominated for generations by the federal government. Welfare policy, in the form of financial transfers to low-income families, had long been deemed a cornerstone of the federal social welfare system. This was attributable in part to the concern that, if left to their own devices, states might engage in a downward bidding war to discourage such families from residing within their boundaries. Through such a "race to the bottom," it was feared, states would continually slash their welfare benefits to such an extent that they would not begin to approach minimum family needs.

Conventional thinking on this issue began to shift in the 1980s and 1990s, ultimately leading to a far-reaching legislative devolution of welfare to the states in 1996.[9] This new legislation continued to provide federal funding for welfare but gave states unprecedented latitude in establishing benefit levels and eligibility rules. It was a direct response to a growing perception that states were proving far more innovative than their federal counterparts in developing welfare programs. The legislation proved contentious at enactment, when it received support from a coalition that included a majority of congressional Republicans and a minority of congressional Democrats before being signed by President Clinton. The new program remains controversial, but its passage illustrates the possibility of a substantial shift in authority in an area traditionally assumed to be the province of the federal government.

Similar forms of devolution have been proposed in virtually every other sphere of domestic policy. The Clinton administration experimented with some elements of this through its "reinventing government" strategies. These offered states far more latitude over policy decisions and use of federal dollars if they would commit to certain performance goals. In environmental policy, this took the form of programs such as the National Environmental Performance Partnership System, which invited states to take the lead in negotiating a new compact between state and nation. Although the NEPPS system was not designed as a climate change program, experience in New Jersey and a few other states illustrated its potential for incorporating greenhouse gas reductions into a larger set of inter-governmental negotiations. Before NEPPS, the federal government experimented with far-reaching devolution of authority to states for such historically nationalized policy areas as management of low-level radioactive waste and cross-boundary transport of ozone.

How realistic is it to give states a significant role in shaping and implementing the next generation of American climate change policy? Is it possible to envision ways for them to take more central, long-term roles in policy development and implementation? The recent proliferation and diversification of state policies to reduce greenhouse gases suggests considerable capacity for innovation. Indeed, virtually any future step that the federal government could conceivably take in coming decades is likely to be borrowed from something already being attempted in one or more states. Although it is indeed difficult to envision an American approach to climate change policy that is exclusively reliant on states, the cases discussed in prior chapters suggests that it might be foolhardy for the federal government to ignore state experience and at some future point try to impose a new national strategy of its own design.

At the same time, many states not only have active policy formation processes but also have substantial populations that generate considerable amounts of greenhouse gases. Many states represent greenhouse gas bases that eclipse individual nations, including a number of European participants in the Kyoto Protocol. In the case of the European Union, individual nations have established varied reduction standards rather than the single reduction target that was created for the United States. Under Kyoto, the EU nations are expected to reduce their greenhouse gases to 8 percent below 1990 levels by 2012. But the targets of individual nations within the EU vary markedly, from reductions of 21 percent in Germany and 12.5 percent in the United Kingdom to limits on increases of 15 percent in Spain and 25 percent in Greece. In between, nations such as Finland and France are simply expected to freeze their 1990 emission levels by 2005 (see table 5-1). In addition to this inter-nation variation, emissions trading is to be facilitated across national boundaries through a flexible "EU bubble" that links all fifteen participating nations.

Ironically, through this formulation the EU has operated like a federated system, acknowledging the differential capacity for greenhouse gas reduction of its various members. In contrast, though the United States operates a federal system, its central government has functioned like a uni-

Table 5-1. *Greenhouse Gas Reduction Targets for European Union Members under the Kyoto Protocol, 1990–2012*

Percent

Member nation	Change from 1990 levels
Austria	−13.0
Belgium	−7.5
Denmark	−21.0
Finland	0
France	0
Germany	−21.0
Greece	+25.0
Ireland	+13.0
Italy	−6.5
Luxembourg	−28.0
Netherlands	−6.0
Portugal	+27.0
Spain	+15.0
Sweden	+4.0
United Kingdom	−12.5

Source: Michael Grubb, with Christiaan Vrolijk and Duncan Brack, *The Kyoto Protocol: A Guide and Assessment* (London: Royal Institute of International Affairs, 1999), p. 123.

tary state in climate change policy development. Neither the Clinton nor Bush administration consulted closely with state counterparts in developing their respective federal strategies for climate change policy, despite states' obvious stake and growing expertise in greenhouse gas reduction. The main exception to this top-down American approach has been federal efforts, most notably, through EPA's State and Local Climate Change Program, to provide financial and technical assistance to states in developing their own reduction initiatives. But these efforts have not extended to serious engagement of states in federal policy development, whether the construction of Kyoto in the Clinton era or the creation of carbon-intensity proposals under Bush.

Development of a more state-centric system of climate change policy faces at least four significant obstacles. Each of these represents a significant limitation on the potential for states to assume more central and enduring roles in climate change policy. These limitations may reflect either internal shortcomings or external opposition to devolution. None is in itself insurmountable, but collectively they give pause to policymakers exploring more devolutionary strategies.

Uneven Performance

State agency officials regularly assumed roles as active policy entrepreneurs in states such as New Jersey, New Hampshire, Oregon, and Wisconsin. They played a central role in developing new policies to reduce greenhouse gases. But not all state capitals provide a comfortable base for launching policy innovation. In Michigan, for example, state officials even attempting such policy development might well be putting their jobs in jeopardy, given that state's aversion to steps that might reduce greenhouse gases. This dichotomy underscores the diverse range of state responses to the challenge of climate change. Indeed, as prior chapters have indicated, many states are hardly ready for prime time. A few, in fact, appear determined to stave off any action on greenhouse gases for as long as possible.

This variation in state responsiveness is not unique to climate change. But it often tends to be overlooked, given the bias toward states and decentralization common in many publications on state-federal relations. Organizations that represent various elements of state government, such as the National Governors' Association and the National Conference of State Legislatures, among others, tend to highlight "success stories" and "best practices" in their publications. This is also true for more specialized organizations that operate in such areas as environmental and energy policy. In turn, leading publications that address state and local government and intergovernmental relations tend to take avowedly pro-state positions. Consequently, discussion of "laggard" or "rogue" states is generally eclipsed by a more standard story line that accentuates state successes.

There is ample evidence to support such an assessment in the case of climate change policy. But states such as Michigan and Colorado have demonstrated outright hostility to the idea of reducing greenhouse gases, whereas other states such as Louisiana and Florida have shown indifference. Of course, additional states outside the sample included in this study fall into these two camps. Indeed, one unusually detailed analysis of overall state capacity in environmental policy acknowledges numerous instances of states that demonstrate levels of leadership and innovation well beyond federal capacity. But the study also notes that "numerous states lay well behind in preparing themselves for a new generation of environmental problems that will require new strategies to be solved."[10] This same variation appears evident in state efforts to reduce greenhouse gases, although there is clearly a steady trend toward greater capacity and innovation.

These shortcomings may only be accentuated by the fact that many states are experiencing their most severe fiscal difficulties in at least a quarter century. States faced a collective shortfall of $100 billion for fiscal year 2003, according to the National Conference of State Legislatures. After an initial period of temporary fixes, many states exhausted one-time options and have been forced to make deep budget cuts, impose large tax increases, or both. Aggregate state fiscal reserves dropped significantly during 2002 and 2003, and a number of states confronted the possibility of downgraded bond ratings, which could increase future costs of borrowing.[11] These fiscal problems could jeopardize many of the positions held by agency-based policy entrepreneurs and threaten the staff teams that have been developed in a number of jurisdictions. Indeed, many state officials with active climate change programs concur that this fiscal threat may imperil the implementation of existing policies and the development of additional ones. Clearly, any longer-term strategy for greenhouse gas reduction that gives states central roles in policy development and implementation will have to address the uneven levels of state commitment and ensure that core fiscal needs are met. Thus far, greenhouse gas reduction programs appear to have fended off efforts to raid their funds for transfer to other functions. In Massachusetts, for example, Republican governor Mitt Romney blocked an effort by the legislature in 2003 to shift $17 million from the state's social-benefit charge to other purposes.

Moreover, many states may face a shortage of political resources, reflecting a substantial turnover of elected officials who worked with policy entrepreneurs to ensure enactment of new climate change policies. The November 2002 elections resulted in change in twenty-four of the fifty governorships. This included selection of replacements for such climate change policy supporters as John Kitzhaber in Oregon (who retired) and Jeanne Shaheen in New Hampshire (who was defeated in a race for the U.S. Senate). In addition, the Bush administration had previously recruited to its cabinet two governors who had strongly supported greenhouse gas reduction in their respective states, Christine Whitman from New Jersey and Tommy Thompson from Wisconsin. Turnover was also substantial at the state legislative level. Nearly one of every four state legislators sworn in during early 2003 was a newcomer, reflecting the highest rate of legislative turnover in a century. This was attributable in part to the impact of legislative term limits, which are in effect in twenty states. However, there has been no indication of state backtracking on climate change owing to this massive change in electoral leadership. In fact, in some instances, such as Michigan and Pennsylvania, there are some indications that the 2002 elections may result in greater political responsiveness to greenhouse gas reduction initiatives.

The Patchwork Quilt of Standards

Historically, analyses of regulatory politics have presumed that regulated parties might prefer decentralized strategies. In theory, giving states the lead role might facilitate the development of a cozy relationship in state capitals between elected officials, state agencies, and the parties they are responsible for overseeing. Under such an arrangement, regulated parties might come to dominate their areas of policy within a given state. This view has changed as states have developed more robust political systems and as more firms operate across state and national boundaries. Many of the state climate change policies introduced in previous chapters reflect this new dynamic, as is evident in their assumption of a lead role in defining policies to reduce greenhouse gases. As their efforts expand, it is possible that there could indeed be a tipping point whereby state-by-state variation of policies and standards creates operational inefficiencies and leads to a call for some form of federal action to establish a uniform policy for the nation.

There is abundant precedent for this type of dynamic in environmental and energy policies, particularly in recent decades, when state activity has increased and private entities "do not feel any particular affinity to the states in which they operate."[12] In the early 1980s, for example, President Reagan refused to establish national standards for energy efficiency in household appliances, although he held that authority under the 1978 Energy Policy and Conservation Act. In response, many states began to develop their own standards, causing concern among appliance manufacturers apprehensive about their ability to market their products nationally, given state regulatory variation. These manufacturers subsequently prodded Congress to act, resulting in the establishment of national standards in the 1987 National Appliance Energy Conservation Act.[13] This process has been repeated in numerous other instances in recent decades, and it is perhaps one of the few forces that can break gridlock in the decisionmaking processes of federal institutions. As the journalist Jonathan Walters has noted, "Business lobbyists do seem to be saying they'd rather deal with the single federal devil they know, than with the hundreds of pesky and unpredictable state and local demons they don't."[14]

This concern has begun to surface as state engagement on climate change has intensified. One corporate executive, Judith Bayer, the director for environmental government affairs at United Technologies, has implored states to avoid creating a "patchwork quilt of government programs . . . in which we operate with potentially conflicting objectives and approaches."[15] Perhaps the clearest manifestation of business concern over state involvement comes from California, where a powerful coalition of private institutions launched an expensive lobbying effort that attempted to block the 2002 legislation to create carbon dioxide emissions standards for motor vehicles. Narrowly defeated in Sacramento, this alliance has now turned to the federal courts in an effort to overturn the legislation as a usurpation of federal regulatory authority. The coalition includes the Alliance of Automobile Manufacturers, the California Chamber of Commerce, and the California Farm Bureau. Members contend that such regulatory intervention is inappropriate for states in a national marketplace for vehicle manufacture. As in other regulatory areas, these plaintiffs would probably prefer no governmental intervention whatsoever. If forced to make a choice, however, many private sector actors would probably prefer one national standard over the possibility of variation by state or region. This type of concern raises potentially significant challenges to any state-driven strategy for reducing greenhouse gases. It also introduces important questions of whether long-term policies could be devised that continue to allow states to take advantage of their particular expertise while not interfering with interstate commerce and other potential constitutional stumbling blocks.

Infrastructure Needs

One of the most active areas for innovation in state climate change policy involves the electricity sector. In particular, the development of renewables portfolio standards and related programs appears to be sparking unprecedented interest in active exploration of renewable-energy alternatives. The experiences of Texas and other states suggest especially favorable prospects for substantial expansion of wind power, leading to potentially substantial reductions in greenhouse gas emissions from electricity generation. Significant advancements in technology appear to make wind power increasingly competitive economically, propelled further by a combination of state mandates and federal tax credits. Moreover, many sections of the United States appear physically well suited for further expansion of wind power.

Realization of the potential of wind power in the United States, however, may depend in large part upon the development of a better infrastructure to move electricity from its point of generation in wind turbines to the point of consumer demand. The states with the greatest potential for wind power, based on physical characteristics, tend to be those located in the geographic center of the nation, essentially North Dakota and Texas and all states located between the two. The current capacity for export of this electricity across state and regional lines, however, is limited at present by the structure of the national electricity transmission system. The existing system has been likened to using "highway maps of the 1930s" to navigate the modern interstate highway system.[16] States such as North Dakota and Texas may literally be impaired in exporting renewable wind energy because of this archaic system. As one analysis has noted, "North Dakota, the nation's windiest state, lacks transmission lines to connect windmill arrays with big-city customers in other regions."[17]

The technical answer to this problem is the development of a more integrated electricity infrastructure. This will entail a much more complex political problem, the siting of substantial new transmission lines. These lines, if sited, will span vast stretches of land and regularly cross state borders. Efforts to create such a system are already proving to be highly controversial, raising fundamental questions about the methods used to secure support for long-distance siting needs. This issue also poses serious intergovernmental challenges. Can states work cooperatively to stitch together a more integrated national system? Does the federal government need to impose a national siting plan? Can any government succeed in overcoming the tremendous public opposition likely to greet any siting plan?

The Federal Energy Regulatory Commission is currently taking a lead role in this matter, but there is no easy resolution in sight.[18] Ironically, Patrick Wood, who was an important figure in the development of the Texas renewables portfolio standard, is now the head of the commission, and as such he is responsible for leading this process. "Nobody wins on transmission issues; it's lose-lose all the way," one state government official explains. In this case, general support for renewables may be trumped by the more immediate concerns about the physical disruption to extensive stretches of land that siting would create. "This kind of an issue really inflames the grass roots, and it cuts across thousands of land sources," acknowledges the leader of a Wisconsin environmental group who is active on energy development issues. "No environmental group wants new lines; neither do most property owners. But the effectiveness of the Wisconsin [renewables portfolio standard] may hinge on getting renewables east to us, and that's going to be a real political problem."

Expansion of renewable energy may also face political resistance within particular state or regional boundaries. Siting of wind turbines has not proved particularly controversial in places such as North Dakota and western Texas, with vast expanses of uninhabited territory. But in response to the transmission difficulties noted above, a number of states with denser populations—and in some cases less suitable physical features for wind power—have attempted to develop their own wind sources. Increasingly, local communities have effectively blocked these proposals, essentially borrowing the opposition tactics so common in other land-use disputes.

Ironically, the so-called NIMBY syndrome may further impair development of wind energy in the very places where demand for renewable electricity is greatest. This phenomenon has been noted in eastern Wisconsin, but it is also evident in other states. In particular, it has emerged as a potential stumbling block in New England's regional efforts to reduce greenhouse gas emissions through generation of wind power. In recent years, a series of wind-power proposals in such states as Massachusetts and Connecticut have been blocked by local opposition concerned about the possible impact on property values, the safety of birds, and such aesthetic considerations as the appearance of tall turbines. "Siting is our biggest issue here for wind power," notes one New England official active in the regional strategy for greenhouse gas reductions. "In states like Massachusetts, you just can't get public approval for wind turbines. In states like Connecticut, there are horrible transmission constraints in the southwest part of the state and a real need to improve that system. But it is one of the wealthiest parts of the nation per capita, and they just won't allow new wind towers or transmission lines."

Federal versus State Authority

The siting of renewable-energy facilities and new transmission lines presents a familiar challenge to federal and state governments, both of which have struggled mightily with analogous cases in the past. But it may be illustrative of a larger question facing any long-term strategy for greenhouse gas reduction that actively involves state governments: Where exactly do state powers yield to the federal government? In turn, how flexible is the federal government prepared to be in allowing active state engagement, given conventional assumptions that an issue such as climate change would involve only federal institutions participating in an international regime?

There is no easy answer to this question, which involves a mixture of political and legal interpretation. Previous attempts to devolve governmental functions from the federal to the state level have foundered for a variety of reasons. These have included federal court interpretations that states were encroaching on federal powers or reluctance by Congress and presidents to actually hand over functions to their state counterparts. Even presidents who have been state governors have often proved ambivalent about yielding federal turf once they are ensconced in Washington, D.C. Moreover, even recent U.S. Supreme Court decisions that appear to constrain some areas of traditional federal jurisdiction hardly constitute a clear doctrine on federalism. In fact, many prominent cases concerning intergovernmental jurisdiction have been settled by close votes in which multiple opinions are issued. So it remains largely unclear just how far states might be allowed to go in defining and implementing climate change policy.

No major political or legal collision has yet occurred between the federal government and states that have attempted major new climate change policy initiatives. But there are at least four areas in which considerable conflict could arise and early examples of potential intergovernmental turf battles have emerged. First, federal legislation may serve to constrain state activity. A series of political actions by the 104th and 105th Congresses clearly had a chilling effect on state policy development. Although congressional actions such as the Byrd-Hagel Resolution in 1997 and a series of legislative riders authored by Representative Joseph Knollenberg (R-Mich.) did not expressly address states, they clearly served to caution states against early actions. In the case of Byrd-Hagel, some states were reluctant to move ahead for fear EPA might try to strike down their actions. "We knew EPA was bound by Byrd-Hagel, and so it was uncertain whether what we were doing would be acceptable," a New Jersey official recalls. "It was ultimately concluded that it only bound the feds from implementing Kyoto, so we just kept going ahead." However, some state officials also note that the aura of Byrd-Hagel served to actively discourage inclusion of climate change issues in NEPPS negotiations, as EPA officials proved reluctant to make any specific references of any sort to greenhouse gases during a crucial period of NEPPS plan development. In addition, the Knollenberg Amendments had a more enduring impact, as they effectively banned the federal government from spending new funds to reduce greenhouse gases. States began to be concerned that, having accepted federal dollars to develop greenhouse gas inventories and action plans, they might be violating federal law if they actually tried to implement programs that would lead to reductions. The federal government never took action to restrict these state expenditures, but the amendments further served to deter state innovation.

Second, active state engagement on climate change does raise fundamental questions of whether new regulatory programs might interfere with interstate commerce. Under the Commerce Clause of the U.S. Constitution, the Supreme Court has long limited state efforts that either burden interstate commerce or discriminate against commerce that originates in another state. This has often led to federal "preemption," whereby state policy initiative is thwarted by what in effect translates to a federal takeover of the policy area in question. Such judicial interpretations have been modified somewhat during the past decade, through a series of Supreme Court decisions that appear to expand state regulatory latitude. As the federalism scholar Michael Greve has noted, "Over the past decade, the Supreme Court has substantially expanded the states' sovereign immunity from federal impositions."[19] Nonetheless, the Commerce Clause could come into play for a constitutional test as states develop more rigorous policies that arguably impair interstate commercial activity.

In one early example of potential legal conflict, Minnesota courts ultimately rejected a challenge to a 1993 state law that called upon the Minnesota Public Utility Commission to examine the range of "environmental costs" associated with various methods of electricity generation.[20] This process was to include carbon dioxide as well as conventional pollutants and was to be used to guide future state planning on electricity. These costs were to be formally considered in proposals to develop new electricity capacity (as was required in Oregon with respect to energy development). Their inclusion offered a clear advantage for renewable sources in any future review of alternatives. Staff from the Minnesota PUC and the Minnesota Pollution Control Agency (PCA) held extensive hearings on the cost assessment process and developed a set of recommendations for each specified pollutant.

In response, Minnesota electricity generators and major industrial consumers of electricity challenged the PUC and PCA interpretations. This opposition was supported, in turn, by North Dakota officials and its coal industry, who feared that the process could reduce the future likelihood of exporting North Dakota coal into Minnesota for its use in electricity generation. Ultimately, this legal challenge was rejected. The administrative law judge Allan Klein established a dollars-per-ton range of environmental costs for the pollutants specified in the 1993 legislation. This included a range of $0.28 to $2.92 per ton as the environmental cost of carbon dioxide. Although the PUC voted in 1997 to accept this interpretation with only minor adjustment, the case serves to underscore the potential for conflict over alleged state infringement of interstate commerce that could ultimately be tested in the federal courts.

The October 2002 decision by the Bush administration's Justice Department to support a legal challenge to California's program to promote low-emission vehicles further illustrates the potential for intergovernmental conflict over what level of government should guide climate change policy. In this instance, the Bush administration did not directly challenge California's new carbon dioxide legislation for vehicles. Instead, it offered a legal brief in support of a challenge by major automakers to the state's program to promote low-emission vehicles, claiming state encroachment on traditional federal regulatory powers. "Regulating fuel economy has always been, and should continue to be, a federal responsibility," according to Chet Lunner of the U.S. Department of Transportation. Insisting that this legislation did not address federal fuel economy standards but rather reflected ongoing California efforts to improve environmental quality, Governor Gray Davis responded, "I'm disappointed that the federal government would intervene with our efforts to protect our air quality."[21]

Third, some states may be increasingly inclined not only to oppose federal preemption initiatives but also to launch their own litigation in attempting to prod some form of federal activity. In recent decades, state attorneys general have become particularly active in developing ambitious litigation strategies, such as their assault on tobacco companies that resulted in massive settlements and revenue gains for states.[22] There are signs that some attorneys general are now turning their sights on areas in which they deem the federal government negligent, including regulation of conventional air pollutants. Particularly relevant to climate change, the attorneys general of Connecticut, Maine, and Massachusetts filed suit in federal court in June 2003 alleging that EPA under President Bush had failed to expand the powers of the Clean Air Act to include regulation of carbon dioxide. This followed a letter sent by seven attorneys general to

President Bush in February 2003 that warned of litigation unless the administration responded. Under the Clean Air Act, EPA is required to review and, where appropriate, revise regulations according to evolving scientific understanding. Contending that carbon dioxide is a pollutant "that causes global warming with its attendant adverse health and environmental impacts," the attorneys general called upon the Bush administration to revise the Clean Air Act to bring carbon dioxide under its regulatory framework.[23] All of the participating attorneys general are Democrats, which may reflect the more partisan and adversarial style of policymaking evident in California that contrasts with the other cases in this book.

A final area of potential legal skirmish involves state efforts to work directly with other national governments. Previous chapters have presented numerous examples of initial exploration of possible collaboration between American states and other nations. New Jersey would like to revisit its earlier involvement with the Netherlands on a series of climate-related issues; Illinois would like to renew initial negotiations with China concerning possible carbon-trading projects; Nebraska would like to explore ways to sell carbon credits from agricultural sequestration to countries that have ratified the Kyoto Protocol. The six states of New England have continued to forge ahead with their formal agreement with the five eastern Canadian provinces, even after Canada ratified Kyoto in November 2002. In fact, they already are exploring ways to expand this regional partnership through the possible addition of other states and provinces as "adjunct" members.

To this point, no legal challenges have been raised to this activity, but these agreements between states and national governments may begin to straddle the boundary of how far states can go in exercising international relations authority traditionally delegated to Washington. The economist Jon Reisman has argued that the entire regional exercise violates constitutional restrictions on state government involvement in cross-jurisdictional treaties and alliances.[24] Indeed, if the Bush administration is concerned about state encroachment on federal authority and is also looking for ways to slow down the pace of state innovation on climate change, it could extend its current litigation strategy in California and attack an array of initiatives, including those that engage states with other nations. One could even envision protracted intergovernmental litigation combat, with the federal government attempting to constrain states from doing so much to reduce greenhouse gases while an expanding alliance of states attempts to force Washington to stop doing so little. There may be more constructive ways, however, to guide future federal-state relations on this issue.

Options for American Climate Change Policy

Long before the Bush administration decided to withdraw the United States from further deliberations over the Kyoto Protocol, this international strategy for greenhouse gas reduction faced innumerable hurdles. The inability to enlist developing nations left gaping holes in overall coverage; many participating nations lacked prior experience with the key policy tools that would be necessary for international emissions trading; and the protocol itself has acquired an ever-expanding series of loopholes that call into question its likely impact.[25] It is, in some ways, reminiscent of other attempts at "policy beyond capacity," whereby an incredibly complex regulatory system is proposed but political support is shaky and the capacity of participating governments to implement it is highly suspect.[26] Analogies to early clean-air programs and Clinton administration proposals for comprehensive health-care reform come to mind, although even these initiatives were modest in comparison with the intended scope of Kyoto.

The Bush action in effect formalized what had been fairly obvious for some time: the U.S. Senate was highly unlikely to ratify Kyoto under any president or any circumstances. At the same time, Kyoto is hardly the end of American involvement in climate change policy. In particular, since the denouement of American involvement in Kyoto, there has emerged a fairly stunning proliferation and diversification of state efforts to create elements of a policy architecture to reduce greenhouse gases. These policies are hardly as dramatic or sweeping as an international regime, but they introduce significant elements of what could evolve into a long-term strategy for the United States to begin to respond to the challenge of climate change. Looking ahead, there are at least four distinct scenarios that indicate what might come next and how these state innovations might influence the next iteration of American climate change policy. Each would offer a way to learn from and build upon recent state experience, but in very different ways. These options reflect both the growing commitment and capacity of states and the likely impediments to state action discussed above.

Federal Replication

There are decades of precedent for basing federal policy on previous state innovations, from Social Security to the Toxics Release Inventory. This reflects a basic dynamic of a federated system of government. As the political scientists Frank Baumgartner and Bryan Jones have noted, "The multiple venues of the states and the federal government sometimes coalesce into a single system of positive feedback, each encouraging the other to enact stronger reforms than might otherwise occur."[27] In the case of climate change, a series of new federal policies in the early 1990s gave states new opportunities, experiences, and resources to begin to contemplate next steps. In more recent years, federal government gridlock and disengagement have further motivated states to accelerate their efforts.

All of this recent state activity offers a policy laboratory for consideration of possible next steps in federal policy. Much important federal energy, environmental, and transportation legislation is due for reauthorization. It has been more than a decade, for example, since the federal government last updated clean-air legislation. Clearly, the best of recent state experience could offer lessons and models for nationwide experimentation. If Massachusetts, New Hampshire, and Oregon can establish regulations to reduce carbon dioxide from electricity generation with little controversy, is it conceivable that the federal government could learn from these initiatives as it contemplates the next round of air-quality policy? If Texas and more than a dozen other states are able to implement renewables portfolio standards and develop significant sources of renewable energy, is the federal government capable of taking such steps? If mandatory carbon dioxide reporting has proved feasible in Wisconsin, is it possible for the federal government to establish a national reporting system?

The 108th Congress is entertaining a variety of proposals that endorse national strategies for multipollutant air controls that include carbon dioxide, mandates for renewable energy, and mandatory greenhouse gas reporting. Bills of this sort have, in fact, been floated in each of the past several Congresses, although none has ever come particularly close to enactment. The impediments to federal adoption of greenhouse gas reduction programs are largely familiar, reflecting the extreme difficulty of enacting any domestic legislation, given partisan and interbranch divides. Indeed, the ongoing futility in enacting any meaningful federal effort to reduce greenhouse gases has been matched by comparable inertia in other salient policy areas.[28]

This kind of problem appears commonly in matters that involve distribution of federal and state powers. Like at least six of his immediate predecessors, President Bush has endorsed the idea of a "new federalism" that would formally devolve greater authority from Washington to state capitals. Translation of broad principles into particulars has proved difficult, however. In fact, state and federal tensions have expanded markedly in the Bush era, with governors from both parties increasingly inclined to blame new federal policies for at least part of their fiscal problems. New federal mandates for homeland security and education have not been accompanied with the dollars necessary for implementation, and some of the recent federal tax cuts have directly reduced state yields from linked tax programs. "White House relations with the governors are as rocky as they've been in decades," according to the political scientist Donald F. Kettl.[29] These growing intergovernmental fissures may increasingly extend to climate change as well, with the Bush administration clearly of two minds as to how to develop federal policy and respond to state initiatives. On the one hand, the administration opposed a national renewables portfolio standard, in large part on the argument that this was a matter best left to states, such as Texas. On the other, the Bush Justice Department has signaled its possible preparedness to attempt to block states from pursuing other greenhouse gas initiatives by arguing that the California approach to reducing vehicle emissions represents an unconstitutional encroachment upon federal powers. Moreover, the Bush administration's EPA complicated matters further in November 2002 by expressing its support for a Massachusetts vehicle program that closely resembles the one in California that has triggered such controversy.

At a minimum, any future federal effort that would attempt to nationalize policies already being implemented in some states needs to be attentive to those experiences and design federal policies accordingly. States could be actively engaged in policy design and given a prominent role in policy development. Nonetheless, the prospects for any near-term breaking of the federal policy logjam appear limited at best. The November 2002 national elections did unify control of the executive and legislative branches under Republican Party control, but that unity seems unlikely to produce greenhouse gas reduction policies. The most receptive venue for climate change in the 107th Congress was the Senate Environment and Public Works Committee, chaired by Vermont independent James Jeffords. The present committee, however, appears far less amenable to serious consideration of greenhouse gas issues, reflecting the views of its new chair, Oklahoma Republican senator James Inhofe. Given this context, major federal legislation, though not impossible, remains unlikely, particularly on an issue such as climate change that appears to elude consensus far more at the national level than in many state capitals.

Diffusion

The diffusion of policy innovations from one state to others has abundant precedent and has been thoroughly studied for decades. States clearly participate in an interactive learning process that leads to policy development. Frequently, a policy idea pioneered by one state is adopted by others, usually beginning with neighboring states but sometimes stretching across the continent. In some instances, there is sufficient activity among multiple states that the federal government attempts to model its own efforts after earlier state experiments. This is the source of some of the concerns about a "patchwork quilt" of regulatory standards discussed above.

States are clearly moving into an accelerated phase of interstate policy diffusion concerning climate change. In some cases, this activity may reflect formal or semiformal networks among state policy entrepreneurs who share ideas and engage in ongoing adaptation of a policy that may have been launched in a single jurisdiction. For example, officials active in the development of the Wisconsin registry program are part of an informal network with officials from approximately twelve other states. These states are beginning to consider the possibility of a standardized registry program that would involve uniform protocols and verification provisions, building on their experience to date. Representatives of these states meet periodically, either in person or by phone conference, to share information and discuss common concerns about registry development. More formal interstate organizations, such as the National Association of State Energy Officials and the Environmental Council of the States, have attempted to increase interstate communication through conferences, publications, and other networking opportunities.

But interstate diffusion can also occur in the absence of networks and even relationships between officials from different states. This may be particularly true given modern technology, which allows easy access to new legislation, legislative proposals, and state agency reports through extensive state government web pages. The diffusion of agricultural carbon sequestration programs appears to follow this type of diffusion pattern. Nebraska's 2000 legislation was essentially replicated in four other states within a year, despite the virtual absence of interstate discussions. "We haven't had much contact with other states on this," a senior official of the Nebraska Department of Natural Resources has noted. "But someone in another state can go online, copy the bill, put their state's name on it, and get a lot of credit for coming up with a new idea. All in a day. It's a pretty good deal."

These types of diffusion reflect a horizontal pattern of states' learning from one another. But it is also possible to envision a vertical pattern, whereby the federal government actively promotes learning and increases the likelihood of diffusion.[30] Federal grants and technical assistance to states through EPA's State and Local Climate Change Program have clearly supported both the development of policy ideas within individual states and the early bridging of those ideas across state boundaries. This program serves, in many respects, as an interstate information broker, not only through grants but also through conferences and publications.

Such horizontal and vertical diffusion are likely to continue in coming years. A significant number of states, for example, have begun to examine the New Hampshire and Massachusetts cases as they contemplate their own versions of "multipollutant" air-quality legislation that includes carbon dioxide. In fact, this activity could continue to accelerate in coming years, compounding the exponential growth rate of state initiatives enacted between 1999 and 2002.

At the same time, states—or even the federal government—might consider formally supporting expanded innovation and possible diffusion. This would entail legislation that expressly authorizes a certain type or range of state-level innovation, either through formal support of a series of pilot projects or through waivers from more conventional requirements in specified types of cases. This approach might help address one enormous concern that hovers over many state policy experiments: How far can states proceed without explicit statutory support? This is a particular concern for policies that are reliant on executive orders and regulatory interpretations. Such decisions are taken by governors and can be reversed or ignored by their successors without the assent of the legislature. In turn, some states are concerned about possible conflicts with existing state and federal statutes, reluctant to "go too far" and thereby face possible rejection by the courts or federal agencies. Indeed, some efforts to use Project XL and NEPPS to advance greenhouse gas reduction plans have foundered on the fact that these were administrative experiments that lacked any statutory foundation.

One model for such empowering legislation would be Minnesota's 1996 Environmental Regulatory Innovations Act. This legislation concluded that "environmental protection could be further enhanced by authorizing innovative advances in environmental regulatory methods."[31] It built on a body of considerable state experimentation in pollution prevention that offered a blend of significant environmental improvement and greater compliance flexibility for regulated parties.[32] Moreover, it was designed to maximize the likelihood that Minnesota could participate actively in federal programs such as Project XL. This put the state firmly behind experimental proposals that might be advanced by state agencies or state-based firms.

This approach is similar to what Wisconsin is attempting through its proposed Green Tier legislation. If enacted, the legislation would provide a formal state embrace of experimentation that would allow its Department of Natural Resources to include greenhouse gas reductions in developing larger regulatory agreements with utilities and other regulated firms. Such legislation, of course, does not prevent the possibility of a court challenge or federal intervention, but it offers state statutory legitimacy to experimentation that may push the boundaries of conventional policy implementation and introduce greenhouse gases into the regulatory decisionmaking process.

The federal government might consider such legislation, particularly if it is serious about encouraging devolutionary strategies. As the legal analyst John Dernbach and his colleagues have written, "In appropriate cases, federal legislation could expressly authorize state use of specific legal tools in a manner that would shield state laws from challenge under the dormant Commerce Clause."[33] The Clinton administration clearly considered such an approach in support of Project XL and NEPPS, and the idea was championed by Senator Joseph Lieberman (D-Conn.). But the administration's relations with Congress were sufficiently sour during the late 1990s that it ultimately did not want to open the door to conflict over any new environmental legislation. Both Project XL and NEPPS clearly suffered from the absence of this statutory grounding. The political scientist Christopher Foreman Jr. has noted the problems that this posed for these Clinton-era experiments, concluding that "a foundation in federal statute law would greatly ease the strain on efforts to extract greater flexibility from the policy process."[34]

As an occasional champion of state authority, however inconsistent in practice, the Bush administration could indeed support such federal legislation and attempt to assemble a cross-partisan coalition to support enactment.[35] Such legislation might give states increased latitude to pursue various greenhouse gas reduction strategies and encourage federal agencies such as EPA and the Department of Energy to work cooperatively with them in attempting pilot projects or even more cross-cutting initiatives. This could be coupled with an expanded program of federal grant support to states to help underwrite the costs of policy development and also to ease possible disincentives to further state innovation caused by the state fiscal crisis.

Going Continental

The diffusion of ideas and even policy collaboration does not have to stop at national boundaries. The burgeoning cooperation on climate change policy between the six states of New England and five provinces of eastern Canada indicates significant possibilities for subnational partnerships on climate change that cross over the 49th Parallel. This evolving zone for greenhouse gas reduction could indeed expand to encompass other jurisdictions, including New Jersey, New York, Maryland, and Pennsylvania, in the near future. For example, in his 2003 State of the State address New York's Republican governor George Pataki called for development of a regional cap on all greenhouse gas emissions generated by power plants in the Northeast. Pataki has also endorsed a regional approach to renewables portfolio standards. Furthermore, there are early suggestions of similar interactions between states and provinces in the Great Lakes Basin and Pacific Northwest. The State of Washington appears particularly active in the Northwest, reflected in its establishment of statewide greenhouse gas reduction goals in its 2003 Washington Sustainability Plan.

Interestingly, the issue of formal North American collaboration never surfaced as a serious possibility during the prolonged negotiations leading to the Kyoto Protocol. Instead, it was Europe that insisted upon formal collaboration beneath a "bubble," consistent with its growing emphasis on federalism-like approaches to public policy. This provision allowed for different levels of pledged reductions among individual European nations but unified them through an agreement to collectively meet a common goal. The bubble, in turn, was intended to allow considerable flexibility among participating nations in achieving their reductions in the least expensive manner possible. Since Kyoto, the EU bubble has become an accepted component in plans to develop an international regime. As the policy analysts John Gummer and Robert Moreland have noted, "The EU's argument was comparable to stating, for example, that in fulfilling its Kyoto commitments, the United States would not place the same burden on the states of West Virginia and Mississippi as it would on California and New York."[36] This approach was highly controversial in the earlier stages, particularly given the significant nation-to-nation differences in reduction commitments (see table 5-1). The United States was particularly active in opposition, but Europe's position ultimately prevailed.[37] This provision served to unify European support for Kyoto, giving it the moral high ground in international deliberations that it has subsequently acquired.

Kyoto set an important precedent in greenhouse gas reduction "for a general provision allowing, in principle, any group of countries to get together and redistribute their commitments."[38] Given this precedent and the beginning of subnational cooperation, to what extent might American states and Canadian provinces expand their collaboration on this issue? Canada generates roughly the same level of greenhouse gas emissions per capita as the United States and pledged a comparable level of reduction under the Kyoto agreements. But like the United States, it has demonstrated considerable resistance to Kyoto ratification. After five years of deliberation, outgoing prime minister Jean Chrétien announced in September 2002 that Canada would proceed with ratification. Ultimately, Canada did ratify Kyoto, despite widespread opposition in the Liberal cabinet and from two of the four other national political parties. In addition, provincial leaders, whatever their views on climate change policy, remain livid over the limited direction they have received from the federal government in defining their potential responsibilities in any future strategies. Canadian federal government plans for achieving their pledged levels of greenhouse gas reductions remain opaque, at best.[39]

Even after Kyoto ratification, Canada has continued to insist that it expects very liberal terms of engagement, including the possibility of securing "clean-energy credits" for export of its natural gas to the United States. This could be especially important in eastern Canada, given substantial new natural gas discoveries off the coasts of Nova Scotia and Newfoundland as well as the existence of a transmission system that makes it easier—and cheaper—to export the gas to the northeastern United States than to provinces in the West. Such credits may be difficult to arrange under Kyoto, a fact that may further disincline Canada from seriously participating. The Canadian system of federalism is so decentralized that individual provinces could, in turn, block implementation of ratification efforts. Thus far, the province of Alberta, which generates nearly one-third of the nation's greenhouse gases, has expressed its vehement opposition to Kyoto and has instead embraced a Bush-like effort that would voluntarily reduce "carbon intensity" in provincial activity that generates greenhouse gases. Regardless of their particular stance on Kyoto, many provinces appear to be following the pattern of the American states, beginning to experiment with greenhouse gas reduction efforts. Much as in the United States, however, there appears to be far more enthusiasm in Canada for developing specific strategies to begin to reduce greenhouse gases than for formal engagement in the Kyoto regime.

But as the New England states and the eastern provinces of Canada have discovered, there may be considerable advantage to a cooperative strategy on climate change, whether or not Canada ever seriously implements its Kyoto pledges. The North American Free Trade Agreement (NAFTA) that involves Canada, the United States, and Mexico, as well as its Canadian-American predecessor agreement, has opened many new avenues for free movement of trade and interjurisdictional cooperation. However, in energy policy, as the political scientist Pietro Nivola has noted, "NAFTA stopped short of achieving a truly open framework for trilateral trade and investment."[40] Much of this restrictiveness has stemmed from concerns over open energy networks expressed by Mexico's government-dominated energy monopolies as well as particular features of American and Canadian energy policies.[41]

Nonetheless, there is already considerable movement of energy across national boundaries and abundant potential for expanded cooperation. Canada already exports more oil to the United States than does Saudi Arabia, and expanded export of natural gas from Canada and Mexico to the United States could facilitate a transition away from coal toward a less carbon-intensive fuel. Remaining borders on the continental electricity grid not only divide American states from one another but also restrict the movement of electricity between Mexico and southern states and across various regions of Canada. Creation of a truly integrated continental infrastructure could allow for freer movement of electricity and could be linked to efforts to expand the use of renewables in all three nations. As Nivola has observed, "All three NAFTA countries could save energy and score big environmental gains if their electric utility grids interfaced seamlessly. . . . But transmission capacity, unsynchronized connections, and insufficient deregulation (open access) in various local systems continue to pose obstacles."[42]

One possible venue for exploration of tripartite cooperation could be the North American Commission on Environmental Cooperation, based in Montreal. The commission was created through a supplemental agreement to NAFTA to ensure that environmental concerns would be incorporated into development of more-open market relations between the three nations. It plays several roles, including dispute resolution and data collection, and has begun to provide support to the New England–eastern Canadian collaboration on climate change. The organization also has substantial expertise in continental energy issues.

Increased continental cooperation could allow for national and subnational differences while taking advantage of mutual opportunities to simultaneously promote greenhouse gas reduction, greater energy efficiency, and economic development. Mexico's status as a developing country exempted it from the Kyoto process, and its overall annual emissions are only about three-fifths of those from Texas (see table 1-2). Nonetheless, the government of President Vicente Fox has expressed interest in engaging in trilateral discussion on energy development and has hinted at possible involvement in a future climate change agreement. Given Mexico's system of federalism, it is possible to envision cooperation at either the national level or through some form of collaboration that involves Mexican states, Canadian provinces, and American states.

Sorting Out American Intergovernmental Relations

Perhaps the most far-reaching policy option for the United States, short of reengagement in the Kyoto Protocol process, would entail using the issue of climate change and greenhouse gas reduction to begin to undertake a more fundamental reinterpretation of state and federal roles in related areas of domestic policy. U.S. presidents from Richard Nixon through George W. Bush have talked about their desire to shift power from Washington back to the states.[43] State governors and legislators have clamored for decades to have more freedom to design their own policies and use federal dollars as they see fit. Countless scholarly books, studies, and commission reports have called for a fundamental shift of power back to the statehouse. Is it conceivable that a global issue such as climate change might best be addressed through a more flexible system that emphasized state needs and opportunities? Could the goal of greenhouse gas reduction be linked to a formal shift of power and resources to the states, as long as they upheld their end and achieved genuine reductions?

One common theme in much of the analysis on the need to reinvent the practice of American federalism is the importance of "accountable devolution."[44] Rather than simply hand over select functions or resources to the states, the federal government instead must find ways to link such transfers with measures of accountability that hold states responsible for their performance. "It's going to take real statesmanship," explains Utah governor Michael Leavitt, one of the nation's most passionate advocates for a new intergovernmental compact. "States need to step up and do some real problem solving; federal officials need to realize that local control isn't just a historical concept, it's a human obsession."[45] Leavitt's selection by President Bush in 2003 to assume control at EPA now provides him with a new platform from which to put his well-known views on federalism into operation.

This type of approach may simply be too much for the current federal political system to contemplate. But how do we best move beyond the inertia at the federal level in all policy areas relevant to climate change? In turn, how do we best harness the creativity and innovation evident in these same areas in a growing number of states? If put to the test, could organizations such as the National Governors' Association, the National Conference of State Legislatures, the Environmental Council of the States, the National Association of State Energy Officials, and the State and Territorial Air Pollution Program Administrators, among others, design a bottom-up system of greenhouse gas reduction that would build on the best of state experience and also work to develop the capacity of states to implement these programs? Or is their continuous call to assume a leadership role in shaping federal policy largely rhetorical, allowing them to mask differential rates of capacity and commitment on issues such as climate change?

Such an approach would begin with the adoption of a bubble for the American states, possibly linked with Canada or Mexico. This would allow some degree of state-by-state variation in total levels of greenhouse gas reduction that would be achieved. These different levels could be negotiated, much as was done in Europe, through a combination of technical and political considerations. States would take the lead in designing this system, beginning with a mechanism to secure state reduction pledges and a set of incentives to facilitate cooperation. These incentives could include supplemental federal transfer dollars and expanded regulatory flexibility to those states that could achieve performance goals for greenhouse gas reductions. This flexibility could also be extended to related regulatory programs, including those for compliance with conventional air pollutants, consistent with the kinds of experiments undertaken in New Jersey and Wisconsin. The degree and intensity of federal oversight would decrease as states demonstrated a capacity to assume leadership roles and deliver on performance commitments. This would be entirely consistent with calls to more carefully calibrate the intensity of federal regulatory oversight to state capacity and performance.[46]

States might also develop a permit system, establishing a fixed number of permits for greenhouse gas emissions that would be allocated among the states.[47] Considerable intrastate and interstate trading would be encouraged as states continued to tailor reduction policies that best suited their economic and energy circumstances. These might well be linked to reform of existing permitting systems for conventional pollutants, part of a multipollutant strategy for utilities and large industries that was designed to look in an integrated way at the total environmental impact of a given firm or facility. Under either approach, some common system for emissions reporting would need to be established, perhaps an extension of the program originated in Wisconsin. A system of monitoring to ensure legitimacy in all trading would also be essential, perhaps one established through formal state and federal collaboration.

States could begin by attempting to achieve an overall level of reduction that met the targets established in the Rio Declaration within a specified period of time. The Kyoto Protocol is now a part of American political history. But the Rio Declaration, which pledged to stabilize greenhouse gas emissions at 1990 levels, was signed by President George H. W. Bush and ratified by the Senate. Its reduction pledges, though not legally binding, would constitute a reasonable starting point for future reduction efforts. Rio carries far less political baggage than Kyoto and may indeed reflect a more reachable level of reductions, given the tremendous growth of American greenhouse gas emissions during the 1990s. Over time, permissible emissions levels could be further reduced, perhaps in concert with neighboring nations. "We all realize that the next generation of environmental legislation needs to develop multipollutant strategies and develop multistate trading structures," explains George Meyer, who served as secretary of the Wisconsin Department of Natural Resources from

1993 to 2001. "If you include greenhouse gases into this and allow for some waivers and flexibility, you could perhaps take advantage of the different situations presented by different states. There are going to be some areas that may not be viable at the state level, like mobile sources. But fixed sources—utilities and industries—are a different story. They are largely under our control already."[48]

This kind of far-reaching transition in regulatory federalism reflects a growing body of analysis that calls for greater emphasis on performance rather than technical adherence to a series of fragmented federal statutes and federal agency commands that may or may not add up to environmental improvements. It is, in many respects, consistent with the basic tenets that led to the creation of the National Environmental Performance Partnership System in the Clinton-Gore administration. The program lacked a statutory foundation and was never fully tested as a model for promoting devolution alongside superior environmental performance. However, it still operates, and a revitalized form could be entered into a legislative package that also incorporated negotiated plans for reduction of greenhouse gases. Indeed, a state-driven approach might begin by building on the most successful aspects of the NEPPS experience and state innovation in climate change policy, working toward a legislative product that would simultaneously devolve authority and establish clear performance goals. "This would be one way to try to bring all of this together," notes Robert Shinn, the former commissioner of the New Jersey Department of Environmental Protection. "Each state tends to get absorbed with one or two air pollutants that seem to be a problem for them. The same goes for EPA. But we need to bring the pieces together, to see how they fit. We just haven't had time to look at it this way before."[49]

There are countless reasons why such an approach is politically highly improbable. Some combination of federal replication, state diffusion, and regionalism are more likely near-term prospects for climate change policy development. Indeed, the capacity of the federal government to seriously engage states on such an issue is questionable. Moreover, after decades of devolution discussion, it is difficult to envision federal authorities, regardless of party and ideology, signing over transfer papers of significant areas of authority.[50] At the same time, it is not clear that a large number of states or their professional organizations would want to assume such authority. For hostile or indifferent states, any such engagement on greenhouse gas reduction would be extremely difficult. Even national organizations, such as the National Governors' Association, the National Conference of State

Legislatures, and the Environmental Council of the States, might well approach this opportunity with caution. Such groups tend to be eager to promote their "best practices" but prefer to avoid the issue of their "worst practices." Indeed, many states seem averse to any mechanism that might evaluate them in comparative fashion, preferring instead a "Lake Wobegon effect," whereby, in the words of the humorist Garrison Keillor, all are essentially deemed "above average."[51]

Nonetheless, it would be refreshing to see states attempt to build on their rapidly evolving record in climate change policy and offer state-based alternatives to their federal counterparts. The very nature of greenhouse gases cuts across numerous policy domains where states have some degree of authority. Many are increasingly demonstrating a creative capacity to use that authority effectively. They recognize the environmental threat posed by climate change and also envision considerable economic development opportunities if they respond in a timely fashion. Comparable policymaking capacity has simply not been evident at the federal level, which continues to thrash about on this issue more than a decade after the introduction of the first legislative infrastructure that began to respond to climate change.

Reconnecting with the World

At some future point, the United States will reconnect on climate change policy with the other nations of the world, both those that have ratified Kyoto and those that have not. That reunion will probably be delayed for some time, while some nations attempt to implement their Kyoto goals with varying degrees of success. Although the United States is no longer part of Kyoto, it is acquiring invaluable skill in developing and implementing policies to reduce greenhouse gases. Indeed, a growing number of American states may well eclipse a number of nations that have ratified Kyoto in their ability both to craft reduction policies and to achieve actual reductions.

The formulation of meaningful policies to reduce greenhouse gas emissions and potentially to influence the direction of climate change will be a long-term process, measurable in generations rather than congressional election cycles. Recent debate at both the national and international levels has been impoverished both by overheated rhetoric and a penchant for getting bogged down in technical quarrels that lead to moral posturing. In contrast, a significant number of American states have begun to chart what that long-term policy process might entail. They are making green-house gas reductions possible and in the process outlining a reasonable set of approaches that could be expanded and might achieve major reductions. They are also proving that these steps can be taken without triggering political warfare. None of this is as dramatic or as sweeping as an international regime that is initialed after all-night bargaining sessions with hundreds of television cameras rolling. It is, in fact, almost stealth-like in its development, having been largely overlooked by the media and national and international climate change elites. Nonetheless, it constitutes a serious and reasonable beginning in an area that has heretofore been best known for its abundance of melodrama and paucity of action.

Notes

Preface

1. Michael Oppenheimer and Robert H. Boyle, *Dead Heat: The Race against the Greenhouse Effect* (Basic Books, 1990). For a more recent variation, see John Firor and Judith E. Jacobson, *The Crowded Greenhouse: Population, Climate Change, and Creating a Sustainable World* (Yale University Press, 2002), chap. 1.

2. Thomas Gale Moore, *Climate of Fear: Why We Shouldn't Worry about Global Warming* (Washington: Cato Institute, 1998).

3. Joseph L. Bast, James M. Taylor, and Jay Lehr, *State Greenhouse Gas Programs: An Economic and Scientific Analysis* (Chicago: Heartland Institute, 2002), p. 2.

4. Thomas Schelling, "The Cost of Controlling Global Warming: Facing the Tradeoffs," *Foreign Affairs,* vol. 76 (November–December 1997), p. 10.

5. David G. Victor, *The Collapse of the Kyoto Protocol and the Struggle to Slow Global Warming* (Princeton University Press, 2001), p. 90.

6. Many of the key components in the Kyoto Protocol were negotiated at the eleventh hour, and it remains unclear whether representatives of individual nations were even awake—much less aware of what was transpiring—when final decisions were approved at a feverish pace at the end of the process. For an overview of key elements of the protocol, see Michael Grubb, with Christiaan Vrolijk and Duncan Brack, *The Kyoto Protocol: A Guide and Assessment* (London: Royal Institute of International Affairs, 1999).

Chapter 1

1. New Jersey Department of Environmental Protection, *Global Climate Change and Greenhouse* Gases (Trenton, 2001), p. 1.

2. Carbon dioxide (CO_2) is the most ubiquitous greenhouse gas and clearly has the greatest impact. However, its effects are intermingled with those of other gases, including methane (CH_4), nitrous oxide (N_2O), hydrofluorocarbons (HFCs), perfluorocarbons (PFCs), and sulfur hexafluoride (SF_6). This book consistently measures greenhouse gases in million metric tons of carbon equivalent (MMTCE), a commonly used metric that reflects the combination of carbon dioxide and the carbon equivalency of the other gases. For the most part, American state strategies have focused primarily on CO_2, although some agriculture and waste management programs have also begun to address methane.

3. National Research Council, *Climate Change Science: An Analysis of Some Key Questions* (Washington: National Academy Press, 2001).

4. James E. Neumann, Gary Yohe, Robert Nicholls, and Michelle Manion, "Sea-Level Rise and Its Effects on Coastal Resources," in Pew Center on Global Climate Change, *Climate Change: Science, Strategies, and Solutions,* ed. Eileen Claussen (Leiden, Netherlands: Brill Academic, 2001), pp. 43–44.

5. As Thomas Gale Moore, of the Cato Institute, has noted, "People winter in Florida, not Duluth." Moore, *Climate of Fear: Why We Shouldn't Worry about Global Warming* (Washington: Cato Institute, 1998), p. 89. As Henry Aaron of the Brookings Institution has noted, "People retire to Florida, not Minnesota." Henry Aaron, "Presidential Address: Seeing through the Fog: Policymaking with Uncertain Forecasts," *Journal of Policy Analysis and Management,* vol. 19 (Spring 2000), p. 200.

6. U.S. Environmental Protection Agency, State and Local Climate Change Program, *Mapping a Cleaner Future* (1998), pp. 4–5; U.S. Global Change Research Program, National Assessment Synthesis Team, *Climate Change Impacts on the United States: The Potential Consequences of Climate Variability and Change* (Cambridge University Press, 2000).

7. A. Denny Ellerman, Paul L. Joskow, Richard Schmalensee, Juan-Pablo Montero, and Elizabeth M. Bailey, *Markets for Clean Air: The U.S. Acid Rain Program* (Cambridge University Press, 2000); Richard E. Cohen, *Washington at Work: Back Rooms and Clean Air,* 2d ed. (New York: Macmillan, 1995); Gary C. Bryner, *Blue Skies, Green Politics: The Clean Air Act of 1990,* rev. ed. (Washington: Congressional Quarterly Press, 1997).

8. Jonathan Walters, "The TEA Generation," *Governing,* May 2002, pp. 70–76.

9. Michael Grubb, with Christiaan Vrolijk and Duncan Brack, *The Kyoto Protocol: A Guide and Assessment* (London: Royal Institute of International Affairs, 1999), p. 37.

10. For a detailed assessment of the Framework Convention, see Social Learning Group, *Learning to Manage Global Environmental Risks,* vol. 1 (MIT Press, 2001); and Michael Grubb, Matthias Koch, and Kay Thomson, *The "Earth Summit" Agreements* (London: Royal Institute of International Affairs, 1993).

11. Al Gore, *Earth in the Balance: Ecology and the Human Spirit* (Boston: Houghton-Mifflin, 1992).

12. Susan Joy Hassol and Randy Udall, "A Climate of Change," *Issues in Science and Technology,* vol. 20 (Spring 2003), pp. 39–46.

13. Matthew L. Wald, "Clinton Seeks to Regulate Common Gas to Clean Air," *New York Times,* November 12, 2000, sec. 1, p. 21.

14. John H. Cushman Jr., "Big Problem, Big Problems: Getting to Work on Global Warming," *New York Times,* December 8, 1998, p. G4.

15. Joby Warrick, "Administration Signs Global Warming Pact," *Washington Post,* November 13, 1998, p. A26.

16. Gore and Bush did have a brief exchange on global climate change in their second debate. Gore stated, "I want to be able to tell my grandson when I'm in my later years that I didn't turn away from the evidence that showed we were doing some serious harm." Bush countered that "I don't think we know the solution to global warming yet. Before we react I think it's best to have the full accounting, full understanding of what's taking place." But the discussion—and most campaign pronouncements on the issue—had little specificity, leaving the plans of both candidates somewhat unclear. See "The Warming Debate," *Washington Post,* October 30, 2000, p. A26.

17. On the limitations of existing federal voluntary programs and the more recent Bush proposals, see David Gardiner and Lisa Jacobson, "Will Voluntary Programs Be Sufficient to Reduce U.S. Greenhouse Gas Emissions?" *Environment,* vol. 44 (October 2002), pp. 25–33.

18. United Nations Conference on Environment and Development, *Agenda 21, Rio Declaration: Forest Principles* (New York, 1992), p. 3.

19. Quoted in Grubb, Koch, and Thomson, *"Earth Summit" Agreements,* p. 62.

20. See David L. Feldman and Catherine A. Wilt, "Climate-change Policy from a Bioregional Perspective," in Michael McGinnis, ed., *Bioregionalism* (New York: Routledge, 1998), p. 140. For an excellent interpretation of these agreements and one of the rare, early analyses of state-level innovation on global climate change, see David L. Feldman and Catherine A. Wilt, "States' Roles in Reducing Global Warming: Achieving International Goals," *World Resources Review,* vol. 6, no. 4 (1994), pp. 570–84.

21. Richard B. Stewart, "Pyramids of Sacrifice?" *Yale Law Journal,* vol. 86 (1977), pp. 1196–1275; William Ophuls, *Ecology and the Politics of Scarcity* (New York: W. H. Freeman, 1977); Matthew A. Crenson, *The Un-Politics of Air Pollution: A Study of Non-Decisionmaking in the Cities* (Johns Hopkins University Press, 1971).

22. Elinor Ostrom, *Governing the Commons: The Evolution of Institutions for Collective Action* (Cambridge University Press, 1990); Elinor Ostrom, Roy Gardner, and James Walker, *Rules, Games, and Common Pool Resources* (University of Michigan Press, 1996); DeWitt John, *Civic Environmentalism: Alternatives to Regulation in States and Communities* (Washington: Congressional Quarterly Press, 1994).

23. National Academy of Public Administration, *Environment.gov* (Washington, 2000); National Academy of Public Administration, *Resolving the Paradox of Environmental Protection: An Agenda for Congress, EPA, and the States* (Washington, 1997); National Academy of Public Administration, *Setting Priorities, Getting Results: A New Direction for EPA* (Washington, 1995); Donald F. Kettl, ed., *Environmental Governance* (Brookings, 2001); Henry N. Butler and Jonathan R. Macey, *Using Federalism to Improve Environmental Policy* (Wash-

ington: American Enterprise Institute Press, 1996); John Ferejohn and Barry Wein-gast, *The New Federalism* (Palo Alto, Calif.: Hoover Institution Press, 1998).

24. Barry G. Rabe, "Power to the States: The Promise and Pitfalls of Decen-tralization," in Norman J. Vig and Michael E. Kraft, eds., *Environmental Policy* (Washington: Congressional Quarterly Press, 2002), chap. 2.

25. Alfred A. Marcus, Donald A. Geffen, and Ken Sexton, *Reinventing Envi-ronmental Regulation: Lessons from Project XL* (Washington: Resources for the Future Press, 2002).

26. Denise Scheberle, *Federalism and Environmental Policy: Trust and the Pol-itics of Implementation,* rev. ed. (Georgetown University Press, 2004), chap. 1.

27. U.S. Congress, Office of Technology Assessment, *Changing by Degrees: Steps to Reduce Greenhouse Gases* (1991), app. B.

28. For analyses of state policy diffusion addressing environmental and energy policy issues with potential relevance to greenhouse gas reduction, see Christopher P. Borick, "Attacking Demand: State Energy Conservation in an Age of Rolling Blackouts," and Dorothy M. Daley, "Understanding the Rise of Voluntary Pro-grams: Exploring Diffusion of Innovation in State Environmental Policy," both papers presented at the Annual Meeting of the American Political Science Associ-ation, Boston, August 29–September 1, 2002.

29. Daniel P. Carpenter, *The Forging of Bureaucratic Autonomy: Reputations, Networks, and Policy Innovation in Executive Agencies, 1862–1928* (Princeton University Press, 2001), pp. 21–23.

30. John Brehm and Scott Gates, *Working, Shirking, and Sabotage: Bureau-cratic Response to a Democratic Public* (University of Michigan Press, 1997), p. 3.

31. Michael Mintrom, *Policy Entrepreneurs and School Choice* (Georgetown University Press, 2000); Lasse Ringius, *Radioactive Waste Disposal at Sea: Pub-lic Ideas, Transnational Policy Entrepreneurs, and Environmental Regimes* (MIT Press, 2001).

32. David G. Victor, *The Collapse of the Kyoto Protocol and the Struggle to Slow Global Warming* (Princeton University Press, 2001); Warwick J. McKibbin and Peter J. Wilcoxen, *Climate Change Policy after Kyoto: Blueprint for a Real-istic Approach* (Brookings, 2002).

33. Paul E. Peterson, *The Price of Federalism* (Brookings, 1995).

34. Bryan D. Jones, *Politics and the Architecture of Choice: Bounded Ratio-nality and Governance* (University of Chicago Press, 2001), p. 103. For other insights on issue framing, see Donald R. Kinder and Lynn M. Sanders, *Divided by Color: Racial Politics and American Ideals* (University of Chicago Press, 1996); Robert J. Shiller, *The New Financial Order: Risk in the Twenty-First Century* (Princeton University Press, 2003); Deborah Lynn Guber, "Environmental Voting in the American States: A Tale of Two Initiatives," *State and Local Government Review,* vol. 33 (Spring 2001), pp. 120–32; and Bryan D. Jones, *Reconceiving Decision-Making in Democratic Politics: Attention, Choice, and Public Policy* (University of Chicago Press, 1994).

35. Shanto Iyengar, *Is Anyone Responsible? How Television Frames Political Issues* (University of Chicago Press, 1991), p. 11.

36. For a recent application of this well-established dichotomy, see Leslie A. Pal and R. Kent Weaver, eds., *The Government Taketh Away: The Politics of Pain in the United States and Canada* (Georgetown University Press, 2003).

37. On the role of focusing events and conditions under which they are—and are not—likely to trigger new policy, see Thomas A. Birkland, *After Disaster: Agenda Setting, Public Policy, and Focusing Events* (Georgetown University Press, 1997).

38. Barry G. Rabe, "Permitting, Prevention, and Integration: Lessons from the States," in Donald F. Kettl, ed., *Environmental Governance* (Brookings, 2001), pp. 14–57.

Chapter 2

1. Peter K. Eisinger, *The Rise of the Entrepreneurial State: States and Local Economic Development Policy in the United States* (University of Wisconsin Press, 1988); Paul E. Peterson, *The Price of Federalism* (Brookings, 1995).

2. John D. Donahue, *Disunited States: What's at Stake as Washington Fades and the States Take the Lead* (Washington: Basic Books, 1997).

3. In one example of a deeply flawed policy, Arizona passed legislation in April 2000 that was designed to encourage citizens to purchase sport-utility vehicles (SUVs) that could operate on either gasoline or cleaner-burning propane. The intent was to provide incremental encouragement for somewhat cleaner engines in the SUV market. However, technical errors in the legislation created loopholes that led to incentive packages of up to $20,000 for each vehicle, an amount so large that many likely purchasers of smaller vehicles with lower overall emissions may have instead chosen an SUV. Moreover, the costs to the state were enormous, reaching 10 percent of Arizona's entire $6 billion budget in 2000 before the legislature stepped in to constrain the incentive program. Jim Carlton, "If You Paid Half Price for That New SUV, You Must Be in Arizona," *Wall Street Journal*, October 26, 2000, p. A1.

4. Even unsuccessful gubernatorial attempts to win the presidency take this approach, perhaps best reflected in Governor Michael Dukakis's emphasis on the so-called Massachusetts miracle in his 1988 campaign against George H. W. Bush.

5. For an overview of this meeting, see Environmental Council of the States, *Report of the ECOS Climate Change Workshop* (Washington, 1998).

6. See Dan Balz, "Gov. Engler Admits No Global Warming toward Gore," *Washington Post*, April 21, 2002, p. A5. Interestingly, despite Engler's enormous political clout and Gore's presence on the ticket, Republican candidates have lost each of the past three presidential elections in Michigan.

7. David Dempsey, *Ruin and Recovery: Michigan's Rise as a Conservation Leader* (University of Michigan Press, 2001), pp. 265–69.

8. Timothy J. Brennan, Karen Palmer, Raymond L. Kopp, Alan J. Krupnick, Vito Stagliano, and Dallas Burtraw, *A Shock to the System: Restructuring America's Electricity Industry* (Washington: Resources for the Future Press, 1996), chap. 7; Dempsey, *Ruin and Recovery*, p. 291.

9. Tom Arrandale, "Global Warming's Brighter Side," *Governing*, April 1999, p. 46.

10. Michigan, House Concurrent Resolution 70 (November 13, 1997), p. 1.

11. Michigan, House Bill 4651 (December 1, 1999), p. 7. In the course of researching and writing this book, I have been asked frequently whether I would be in violation of this law if it had been enacted, given that the University of Michigan is a public sector institution in the State of Michigan. I have been advised that I fall outside the scope of this law and sincerely hope that this interpretation is correct.

12. Balz, "Gov. Engler Admits No Global Warming toward Gore."

13. Stranded costs raise the question of whether state governments or electricity consumers should compensate utilities for investments, such as new power plants, made by utilities with state approval in a regulated environment. These investments might not have been made, it is argued, had utilities known they would be implemented in a deregulated marketplace. Recom Applied Solutions, *Global Climate Change Strategic Plan for Colorado* (Longmont, Colo., 1997). For a more detailed discussion of stranded costs, see Timothy J. Brennan, Karen L. Palmer, and Salvador A. Martinez, *Alternating Currents: Electricity Markets and Public Policy* (Washington: Resources for the Future Press, 2002), chap. 14.

14. Colorado Department of Public Health and Environment, *Climate Change and Colorado* (Denver, 1998).

15. Colorado, Senate Joint Resolution 98-023 (May 4, 1998), p. 2.

16. See, for example, John Fischbach and Lucinda Smith, "Climate Protection: Fort Collins Likes the Idea," *Public Management*, August 2000, pp. 6–10. The council now has forty-two city participants from twenty-six states, all of which are pursuing a wide range of environmental sustainability initiatives, a number of which could reduce greenhouse gas releases. Few of these have been formally linked to state government actions, although there is considerable potential for this in the future. This study does not examine local government strategies in particular detail, although it recognizes local government as a potentially important supplemental contributor to long-term bottom-up approaches that address climate change as well as other environmental issues. For an excellent analysis of the rapidly expanding role of local governments in initiating environmental innovations, many of them highly relevant to climate change, see Kent E. Portney, *Taking Sustainable Cities Seriously: Economic Development, the Environment, and Quality of Life in American Cities* (MIT Press, 2003).

17. U.S. Environmental Protection Agency, State and Local Climate Change Program, *Partnerships and Progress: 2001 Progress Report* (2002), p. 22.

18. For a useful overview of these earlier programs, see DeWitt John, *Civic Environmentalism: Alternatives to Regulation in States and Communities* (Washington: Congressional Quarterly Press, 1994), pp. 204–49.

19. James E. Neumann, Gary Yohe, Robert Nicholls, and Michelle Manion, "Sea-Level Rise and Its Effects on Coastal Resources," in Pew Center on Global Climate Change, *Climate Change: Science, Strategies, and Solutions*, ed. Eileen Claussen (Leiden, Netherlands: Brill Academic, 2001), pp. 49–52.

20. Ibid., p. 49.

21. Virtus Energy Research Associates, *Texas Energy: Past, Present, Future* (Austin, Tex., 1998), pp. 46–51; Ed Smeloff and Peter Asmus, *Reinventing Electric Utilities* (Washington: Island Press, 1997), p. 134.

22. Andrew C. Revkin, "Texas Takes Step on Warming; Some See Shift in Bush's Position," *New York Times*, August 24, 2000, p. A20.

23. Howard Gruenspecht and Paul Portney, "Energy and Environmental Policy," in Henry J. Aaron, James M. Lindsay, and Pietro S. Nivola, eds., *Agenda for the Nation* (Brookings, 2003), p. 268.

24. William T. Gormley Jr., *The Politics of Public Utility Regulation* (University of Pittsburgh Press, 1983); Richard F. Hirsh, *Power Loss: The Origins of Deregulation and Restructuring in the American Electric Utility System* (MIT Press, 1999).

25. Texas Public Utility Regulatory Act of 1999, sec. 39.904, p. 56.

26. National Association of State Energy Officials, *Energy Efficiency and Renewables Sources: A Primer* (Washington, 1998), p. 46.

27. Texas Public Utility Regulatory Act of 1999, sec. 39.904, p. 56.

28. Alexei Barrionuevo and Russell Gold, "Texas May Face a Glut of Electricity, but That Won't Aid Rest of U.S.," *Wall Street Journal*, May 7, 2001, p. A1.

29. For a useful overview of the evolution of electricity policy in Texas in the 1970s and 1980s, see Smeloff and Asmus, *Reinventing Electric Utilities*, pp. 132–33.

30. Virtus Energy Research Associates, *Texas Energy*, p. 7.

31. Ibid., p. 19.

32. James Fishkin, *Democracy and Deliberation: New Directions for Democratic Reform* (Yale University Press, 1991); James Fishkin, *The Voice of the People: Public Opinion and Democracy* (Yale University Press, 1995).

33. A thorough analysis of the various polling sessions concludes that "deliberation leads to sobered recognition that current renewable technology cannot meet a very high proportion of electricity needs in the short term. But we have also seen that it tends to strengthen support for including renewable energy as part of a long-term strategy." Robert C. Luskin, James S. Fishkin, and Dennis L. Plane, "Deliberative Polling and Policy Outcomes: Electric Utility Issues in Texas," paper presented at the Annual Meeting of the Midwest Political Science Association, Chicago, April 1–3, 1999, p. 8.

34. Thaddeus Herrick, "The New Texas Wind Rush," *Wall Street Journal*, September 23, 2002, p. B1.

35. Virtus Energy Research Associates, *Texas Energy*, pp. 139–61.

36. Ryan Wiser and Ole Langniss, *The Renewables Portfolio Standard in Texas: An Early Assessment* (Berkeley, Calif.: Ernest Orlando Lawrence Berkeley National Laboratory, 2001), p. 4.

37. Anthony Downs, "How Real Are Transit Gains?" *Governing*, March 2002, p. 60.

38. Georgia, House Resolution 441 (March 24, 1999), p. 2.

39. Robert D. Bullard, Glenn S. Johnson, and Angel O. Torres, *Sprawl City: Race, Politics, and Planning in Atlanta* (Washington: Island Press, 2000); Clarence N. Stone, *Regime Politics: Governing Atlanta* (University Press of Kansas, 1989).

40. Research conducted by the Southern Coalition for Advanced Transportation also indicates that the Clean Air Campaign has had high public visibility and has contributed to attitudinal change that has increased receptivity to alternative modes of transportation. Southern Coalition for Advanced Transportation, "Evaluation of the Effectiveness of Programs Contained in the Framework for Cooperation to Reduce Traffic Congestion and Improve Air Quality, Phase Two," report prepared for the Georgia Department of Transportation, Project 9906 (Atlanta: Southern Coalition for Advanced Transportation, 2002), chap. 3, p. 76.

41. Richard H. Adams, Brian H. Hurd, and John Reilly, "Impacts on the U.S. Agricultural Sector," Pew Center on Global Climate Change, *Climate Change: Science, Strategies, and Solutions,* ed. Eileen Claussen (Leiden, Netherlands: Brill Academic, 2001), p. 39.

42. For a concise discussion of the global-warming potential of various greenhouse gases, see Warwick J. McKibbin and Peter J. Wilcoxen, *Climate Change Policy after Kyoto: Blueprint for a Realistic Approach* (Brookings, 2002), pp. 12–14.

43. Adams, Hurd, and Reilly, "Impacts on the U.S. Agricultural Sector," p. 40.

44. Nebraska Department of Natural Resources, *Carbon Sequestration, Greenhouse Gas Emissions, and Nebraska Agriculture: Background and Potential* (Lincoln, 2001), pp. v–vi.

45. John Dernbach and the Widener University Law School Seminar on Global Warming, "Moving the Climate Change Debate from Models to Proposed Legislation: Lessons from State Experience," *Environmental Law Reporter,* vol. 30 (November 2000), p. 10970.

46. Natural Resource Conservation Service, *Quantifying the Change in Greenhouse Gas Emissions due to Natural Resource Conservation Practice Application in Nebraska: The Nebraska Carbon Storage Project* (Lincoln, 2002), p. vi.

47. On the unique features of the Nebraska legislature and their impact on policy formation in the state, see Gerald C. Wright and Brian F. Schaffner, "The Influence of Party: Evidence from the State Legislature," *American Political Science Review,* vol. 96 (June 2002), pp. 367–80.

48. Nebraska Department of Natural Resources, *Carbon Sequestration, Greenhouse Gas Emissions, and Nebraska Agriculture*; Gary D. Lynne and Colby E. Kruse, *Potential for Market System/Carbon Trading,* report prepared for the Nebraska Department of Natural Resources (Lincoln, Neb.: University of Nebraska Public Policy Center, 2001).

49. Nebraska Department of Natural Resources, *Carbon Sequestration, Greenhouse Gas Emissions, and Nebraska Agriculture,* p. viii.

50. Illinois, House Bill 842 (August 7, 2001); North Dakota, Senate Concurrent Resolution 4043 (March 23, 2001); Oklahoma, House Bill 1192 (April 16, 2001); Wyoming, House Bill 0047 (March 1, 2001).

51. Cyd Janssen, interview by author, Gordon, Nebraska, April 19, 2002. On Kerrey's interest in this issue, see Joby Warrick, "Cultivating Farms to Soak Up Greenhouse Gas," *Washington Post,* November 23, 1998, p. A3.

52. Janssen, interview.

53. Merton L. Dierks, *Introducer's Statement of Intent: LB 804, January 30, 2001,* 97th Nebraska Legislature, 1st sess. (Lincoln, 2001).

Chapter 3

1. Jeanne Shaheen, "State, Local, and Corporate Climate Actions Enhance Quality of Life," in Pew Center on Global Climate Change, *Climate Change: Science, Strategies, and Solutions,* ed. Eileen Claussen (Leiden, Netherlands: Brill Academic, 2001), p. 281.

2. Remarks by Robert C. Shinn Jr., in Environmental Council of the States, *Report of the ECOS Climate Change Workshop* (Washington, 1998), p. 49.

3. David Vogel, *Trading Up: Consumer and Environmental Regulation in a Global Economy* (Harvard University Press, 1995); William R. Lowry, *The Dimensions of Federalism: State Governments and Pollution Control Policies,* rev. ed. (Duke University Press, 1996), chap. 4.

4. "Swift Unveils Nation's Toughest Power Plant Regulations," press release, Commonwealth of Massachusetts, April 23, 2001.

5. David Biello, "New Hampshire Imposes CO_2 Cap," *Environmental Finance,* vol. 3 (May 2002), p. 11.

6. Tom Arrandale, "The Pollution Puzzle," *Governing,* August 2002, p. 22.

7. According to New Hampshire Senate Bill 159, which established the state's greenhouse gas reduction inventory, "For past air pollution reduction programs, the federal government has not always afforded New Hampshire sources appropriate consideration for emissions reductions made prior to the implementation of such programs." New Hampshire, Senate Bill 159 (July 6, 1999), p. 1.

8. Jeffrey C. MacGillivray and Kenneth A. Colburn, "The Industry-Average Performance System for Air Pollution Control: A Competitive, Self-Governing Air Pollution Control System," in *Ideas for New Hampshire: 1997 Better Government Competition Winners* (Concord, N.H.: Josiah Bartlett Center for Public Policy, 1997), pp. 1–9. In its form as a legislative proposal, see New Hampshire, House Bill 724-FN (August 10, 1999).

9. New Hampshire Department of Environmental Services, *The Climate Change Challenge: Actions New Hampshire Can Take to Reduce Greenhouse Gas Emissions* (Concord, 2001), pp. 11–12.

10. State and Territorial Air Pollution Program Administrators and the Association of Local Air Pollution Control Officials, *Reducing Greenhouse Gases and Air Pollution: A Menu of Harmonized Options* (Washington, 1998).

11. Kenneth Colburn, quoted in U.S. Environmental Protection Agency, State and Local Climate Change Program, "Integrating Priorities in New Hampshire," in *Solutions and Successes: A Report on the 2000 EPA State and Local Climate Change Program Partners' Conference* (U.S. Environmental Protection Agency, 2000), p. 12.

12. New Hampshire Department of Environmental Services, *Climate Change Challenge,* p. 59.

13. Shaheen, "State, Local, and Corporate Climate Actions Enhance Quality of Life," p. 282.

14. Biello, "New Hampshire Imposes a CO_2 Cap," p. 11.

15. Quoted in Jim Graham, "State First to Take on Global Warming," *Concord (N.H.) Monitor,* April 19, 2002.

16. "Governor Jeanne Shaheen Media Release," press release, State of New Hampshire, May 9, 2002.

17. New Hampshire Department of Environmental Services, *Climate Change Challenge*, p. 29.

18. On the evolution of the siting of energy plants over the past half century, see Barry G. Rabe, "Siting Nuclear Waste," in Leslie A. Pal and R. Kent Weaver, eds., *The Government Taketh Away: The Politics of Pain in the United States and Canada* (Georgetown University Press, 2003), pp. 195–232.

19. William R. Lowry, *Dam Politics: Restoring American Rivers* (Georgetown University Press, 2003).

20. Renewable Resource Institute, *The State of the States Report: Assessing the Capacity of States to Achieve Sustainable Development through Green Planning* (San Francisco, 2000), p. 40.

21. National Association of State Energy Officials, *Energy Efficiency and Renewables Sources: A Primer* (Washington, 1998), p. 27. On the evolution of a number of Oregon energy programs, see Ed Smeloff and Peter Asmus, *Reinventing Electric Utilities* (Washington: Island Press, 1997), chap. 5.

22. National Association of State Energy Officials, "Oregon Climate Trust," in *State Energy Cases* (Washington, 2002), p. 9.

23. Sam Sadler, "Oregon," appendix to Environmental Council of the States, *Report of the ECOS Climate Change Workshop* (Washington, 1998), p. 79.

24. Philip H. Carver, Sam Sadler, Mark C. Trexler, and Laura H. Kosloff, "The Changing World of Climate Change: Oregon Leads the States," *Electricity Journal*, vol. 10 (May 1997), pp. 53–63.

25. Oregon, House Bill 3283 (June 26, 1997).

26. U.S. Environmental Protection Agency, "Oregon Switches to Cleaner Power," in *Climate Change Solutions* (2000), p. 1.

27. Oregon Office of Energy, *Oregon Climate Trust* (Salem, 1998).

28. Matthew Brown, "Innovative State Programs: New Jersey and Oregon Take the Lead," in Pew Center on Global Climate Change, *Climate Change: Science, Strategies, and Solutions*, ed. Eileen Claussen (Leiden, Netherlands: Brill Academic, 2001), p. 306.

29. Christopher Swope, "Volt Revolt," *Governing*, April 2002, pp. 31–33.

30. John Dernbach and the Widener University Law School Seminar on Global Warming, "Moving the Climate Change Debate from Models to Proposed Legislation: Lessons from State Experience," *Environmental Law Reporter*, vol. 30 (November 2000), p. 10968.

31. Oregon, House Bill 2200 (July 6, 2001).

32. U.S. Environmental Protection Agency, State and Local Climate Change Program, *Mapping a Cleaner Future* (1998), p. 15.

33. For an extensive discussion of Portland's active engagement in a wide range of environmental protection initiatives, see Kent E. Portney, *Taking Sustainable Cities Seriously: Economic Development, the Environment, and Quality of Life in American Cities* (MIT Press, 2003).

34. U.S. Environmental Protection Agency, State and Local Climate Change Program, *Partnerships and Progress: 2001 Progress Report* (2002), pp. 25–26.

35. Andrew Caffrey and Robert Gavin, "Power-Plant Momentum Runs Out of Energy," *Wall Street Journal*, May 13, 2002, p. A2.

36. David G. Victor, *The Collapse of the Kyoto Protocol and the Struggle to Slow Global Warming* (Princeton University Press, 2001), p. 112; see chap. 3 for an extended discussion of the challenges of measuring, monitoring, and enforcing any international agreement on greenhouse gas reduction. On the considerable uncertainties for measurement of specific greenhouse gases, see Dean Anderson and Michael Grubb, eds., *Controlling Carbon and Sulphur: Joint Implementation and Trading Initiatives* (London: Royal Institute of International Affairs, 1997), p. 204.

37. William T. Gormley Jr. and David L. Weimer, *Organizational Report Cards* (Harvard University Press, 1999).

38. Barry G. Rabe, "Federalism and Entrepreneurship: Explaining American and Canadian Innovation in Pollution Prevention and Regulatory Integration," *Policy Studies Journal*, vol. 27, no. 2 (1999), pp. 288–306.

39. Mary Graham, "Is Sunshine the Best Disinfectant?" *Brookings Review*, vol. 20 (Spring 2002), p. 18.

40. Wisconsin Department of Natural Resources, *Air Contaminant Emission Inventory Reporting Requirements*, NR 438.03 (May 1993).

41. Mark Stephen, "Environmental Information Disclosure Programs: They Work, but Why?" *Social Science Quarterly*, vol. 83 (March 2002), pp. 190–205; Mary Graham, *Democracy by Disclosure: The Rise of Technopopulism* (Brookings, 2002).

42. Wisconsin Department of Natural Resources and the Wisconsin Climate Change Committee, *Wisconsin Climate Change Action Plan: Framework for Climate Change Action* (Madison, 1998). For a sampling of earlier analytical work that facilitated this report, see Wisconsin Department of Natural Resources and Public Service Commission of Wisconsin, *Wisconsin Greenhouse Gas Emission Reduction Cost Study*, vols. 1 and 2 (Madison, 1996).

43. Wisconsin Department of Natural Resources and the Wisconsin Climate Change Committee, *Wisconsin Climate Change Action Plan*, p. 19.

44. Lowry, *Dimensions of Federalism*, chap. 2.

45. George Meyer, interview by author, Madison, Wisconsin, April 4, 2002.

46. Quoted in Joel Eskovitz, "Wisconsin's Approach to Global Warming Is Largely Voluntary," Associated Press State and Local Wire, August 5, 2002.

47. George Meyer, *A Green Tier for Greater Environmental Protection* (Madison: Wisconsin Department of Natural Resources, 1999); quoted in Graham K. Wilson, "Importing Cooperation," paper presented at the Annual Meeting of the Midwest Political Science Association, Chicago, April 4–6, 2002, p. 29.

48. The Netherlands in particular has been singled out in many scholarly analyses as a model of regulatory innovation and effectiveness. For an overview, see Graham K. Wilson, "Regulatory Reform on the World Stage," in Donald F. Kettl, ed., *Environmental Governance* (Brookings, 2001), pp. 118–45.

49. Bill Clinton and Al Gore, *Reinventing Environmental Regulation* (March 16, 1995) (www.govinfo.library.unt.edu/npr/library/rsreport/251a.html [May 19, 2002]).

50. Wilson, "Importing Cooperation," p. 21.

51. Terry Davies, with Aracely Alicea, Robert Hersh, and Ruth Greenspan Bell, *Reforming Permitting* (Washington: Resources for the Future Press, 2002), p. 64.

52. Ibid., p. 67.

53. Meyer, interview.

54. Wilson, "Importing Cooperation," p. 7.

55. Ibid., p. 30.

56. Jeff Deyette, Steve Clemmer, and Deborah Donovan, *Plugging in Renewable Energy: Grading the States* (Cambridge, Mass.: Union of Concerned Scientists, 2003), chaps. 5–6.

Chapter 4

1. The state is also hammered regularly in barbs that attack the perceived quality of life in New Jersey or the basic competence of its citizens. The journalist William McGurn has noted Governor Whitman's "well-publicized tussles with advertisers who treat New Jersey as one big joke—most recently with Time Warner after the company advertised a cable service 'so simple even someone from Jersey can use it.'" William McGurn, "Name That Tune: New Jersey's Song of Songs," *Wall Street Journal*, December 1, 2000, p. W19.

2. Christine Todd Whitman, quoted in New Jersey Department of Environmental Protection, *Global Climate Change and Greenhouse Gases* (Trenton, 2001), p. 1.

3. Barry G. Rabe, "Permitting, Prevention, and Integration: Lessons from the States," in Donald F. Kettl, ed., *Environmental Governance* (Brookings, 2001), pp. 14–57.

4. New Jersey Executive Department, Executive Order No. 219, February 1989.

5. U.S. Environmental Protection Agency, State and Local Climate Change Program, *Mapping a Cleaner Future* (1998), p. 11.

6. Peter Sugarman, *Sea Level Rise in New Jersey* (Trenton: New Jersey Geological Survey, 1998), p. 2.

7. Robert C. Shinn Jr., interview by author, Hainesport, New Jersey, April 11, 2002.

8. Ibid.

9. Ibid.

10. Ibid.

11. New Jersey Department of Environmental Protection, Administrative Order No. 1998-09, March 17, 1998; reprinted in New Jersey Department of Environmental Protection, *New Jersey Sustainability Greenhouse Gas Action Plan* (Trenton, 2002), app. C, pp. A8–A10.

12. Ibid., app. C, p. A10.

13. Robert C. Shinn Jr. and Matt Polsky, "The New Jersey Department of Environmental Protection's Non-Traditional Role in Promoting Sustainable Development Internationally," *Seton Hall Journal of Diplomacy and International Relations*, vol. 3 (Summer–Fall 2002), pp. 93–103.

14. Quoted in John J. Fialka, "States Are Stepping In to Reduce Levels of Carbon Dioxide," *Wall Street Journal*, September 11, 2001, p. A28.

15. Paul de Jongh, *Our Common Journey: A Pioneering Approach to Environmental Management* (New York: St. Martin's, 1999).

16. Graham K. Wilson, "Regulatory Reform on the World Stage," in Donald F. Kettl, ed., *Environmental Governance* (Brookings, 2001), p. 128.

17. Some of the corporate participants in the New Jersey covenant system have also become involved in the Partnership for Climate Action, which involves corporations' reporting their greenhouse gas reductions and then being monitored by the environmental group Environmental Defense. See Peter Behr, "Big Firms Join to Share Greenhouse Gas Cuts," *Washington Post*, October 18, 2000, p. E3. For a more general overview of corporate innovations on greenhouse gas reductions, see Joseph Romm, *Cool Companies* (Washington: Island Press, 1999).

18. Quoted in Kirk Johnson, "Global Warming Moves from Impassioned Words to Modest Deeds," *New York Times*, November 19, 2000, sec. 1, p. 47.

19. New Jersey Department of Environmental Protection, *Covenant between the New Jersey Department of Environmental Protection and PSEG Fossil, L.L.C.* (Trenton, January 11, 2002).

20. Lynn F. Stiles and Alice Gitchell, "Determination of Reduction of Carbon Dioxide Emissions Due to Geothermal Heat Pump Installations," paper prepared for the New Jersey Board of Public Utilities, November 23, 1999.

21. "PEQ and NJ Department of Environmental Protection Sign Historic Covenant of Sustainability," *PEQ Views*, vol. 2 (Summer 2001), p. 1; Kirk Johnson, "Religious Group Sees Moral Choice in the Monthly Light Bill," *New York Times*, July 10, 2000, p. B1.

22. New Jersey Interagency Sustainability Working Group, *Governing with the Future in Mind: Working Together to Enhance New Jersey's Sustainability and Quality of Life* (Trenton, 2001), p. 25.

23. For a comparison of different state electricity policies, see Steve Clemmer, Ben Paulos, and Alan Nogee, *Clean Power Surge: Ranking the States* (Cambridge, Mass.: Union of Concerned Scientists, 2000).

24. New Jersey Department of Environmental Protection, *New Jersey Sustainability Greenhouse Gas Action Plan: 2001 Addendum* (Trenton, 2002), p. 6.

25. New Jersey Clean Energy Collaborative, *Comprehensive Resource Analysis Report Submitted to the New Jersey Board of Public Utilities* (Trenton: New Jersey Board of Public Utilities, 2002).

26. New Jersey, *Electric Discount and Energy Competition Act*, Public Law 1999 (February 9, 1999), chap. 23, p. 21.

27. Ibid., p. 41.

28. New Jersey Department of Environmental Protection, *New Jersey Sustainability Greenhouse Gas Action Plan*, p. 40.

29. Barry G. Rabe, "Environmental Regulation in New Jersey: Innovations and Limitations," *Publius: The Journal of Federalism*, vol. 21 (Winter 1991), pp. 91–94.

30. Remarks by Robert C. Shinn Jr., in Environmental Council of the States, *Report of the ECOS Climate Change Workshop* (Washington, 1998), p. 50.

31. New Jersey Department of Environmental Protection, *New Jersey Sustainability Greenhouse Gas Action Plan* (Trenton, 1999), p. 41.

32. Ibid., pp. 42–43.

33. New Jersey Interagency Sustainability Working Group, *Governing with the Future in Mind,* p. 23.

34. Denise Scheberle, *Federalism and Environmental Policy: Trust and the Politics of Implementation,* rev. ed. (Georgetown University Press, 2004), chap. 1.

35. National Academy of Public Administration, *Resolving the Paradox of Environmental Protection: An Agenda for Congress* (Washington, 1997), p. 166. The unusually comprehensive nature of the New Jersey NEPPS proposals is also highlighted in Eric Siy, Leo Koziol, and Darcy Rollins, *The State of the State Report* (San Francisco: Renewable Resource Institute, 2001), p. 45.

36. National Academy of Public Administration, *Resolving the Paradox of Environmental Protection,* p. 155.

37. Rabe, "Permitting, Prevention, and Integration: Lessons from the States."

38. New Jersey, *Letter of Intent between the Ministry of Housing, Spatial Planning, and the Environment, The Netherlands, and the Department of Environmental Protection, the State of New Jersey, June 5, 1998;* reprinted in New Jersey Department of Environmental Protection, *New Jersey Sustainability Greenhouse Gas Action Plan,* app. O, pp. A11–A12.

39. Steve Strunsky, "Pact between New Jersey and Netherlands May Mean Cleaner Air for State," *New York Times,* December 24, 1999, p. B6.

40. Shinn, interview.

41. Laura Mansnerus, "New Jersey Intends to End Incentive Plan on Pollution," *New York Times,* September 18, 2002, p. B1.

42. As the state policy analyst Matthew Brown has noted, the state climate change initiatives appear to have broad support from the private sector and elected officials. He contends, however, that "the key to its success in the future will be in finding ways to continue that support through future administrations." Matthew Brown, "Innovative State Programs: New Jersey and Oregon Take the Lead on Climate Change," in Pew Center on Global Climate Change, *Climate Change: Science, Strategies, and Solutions,* ed. Eileen Claussen (Leiden, Netherlands: Brill Academic, 2001), p. 302.

43. Tracey L. Regan, "N.J. Lags in Complying with Clean Air Rules," *Trenton Times,* April 11, 2002, p. A13.

44. Stephen R. Dujack, "The Integrator," *Environmental Forum,* vol. 19 (July–August 2002), pp. 44–47.

45. Bradley M. Campbell, interview by author, New Jersey Department of Environmental Protection, Trenton, April 11, 2002.

46. Rabe, "Environmental Regulation in New Jersey," pp. 98–99; Philip J. Landrigan, Louise A. Halper, and Ellen K. Silbergeld, "Toxic Air Pollution across a State Line: Implications for the Siting of Resource Recovery Facilities," *Journal of Public Health Policy,* vol. 10 (Autumn 1989), pp. 309–23.

47. Committee on the Environment and the Northeast International Committee on Energy, *Climate Change Action Plan 2001,* paper presented at the twenty-sixth annual Conference of New England Governors and Eastern Canadian Premiers, Westbrook, Connecticut, August 26–28, 2001, p. 7.

48. Ibid.

49. Shinn and Polsky, "New Jersey Department of Environmental Protection's Non-Traditional Role in Promoting Sustainable Development Internationally," p. 100.

50. National Academy of Public Administration, *Environment.gov* (Washington, 2000), p. 153.

51. Christopher Swope, "McGreevey's Magic Map," *Governing*, May 2003, pp. 43–50.

52. Shinn, interview.

53. The address, along with other governors' State of the State addresses, is available at www.stateline.org/stateline (October 10, 2003).

54. Gina Keating, "California to Regulate Car Emissions," *National Post*, July 23, 2002, p. 12; Miguel Bustillo, "Exhaust Legislation May Hit a Red Light," *Los Angeles Times*, May 21, 2002, pt. 1, p. 1.

55. Donald F. Kettl, "Sacramento Rules," *Governing*, December 2002, p. 14.

56. Carl Ingram, "Senate Votes to Require Cleaner-Running Cars, Light Trucks," *Los Angeles Times*, May 3, 2002, pt. 2, p. 8.

57. Gary Polakavic and Miguel Bustillo, "Assembly Passes Bill to Control Emissions of Greenhouse Gases," *Los Angeles Times*, January 31, 2002, pt. 2, p. 7.

58. William Booth, "Calif. Takes Lead on Auto Emissions; Gov. Davis to Sign Law on Pollution That May Affect All U.S. Drivers," *Washington Post*, July 22, 2002, p. A1; Gray Davis, "California Takes on Air Pollution," *Washington Post*, July 22, 2002, p. A15.

59. Jeffrey Ball, "U.S. Joins Fight against California Clean-Air Effort," *Wall Street Journal*, October 10, 2002, p. A2.

60. de Jongh, *Our Common Journey.*

Chapter 5

1. Robert D. Putnam, "Diplomacy and Domestic Politics: The Logic of Two-Level Games," *International Organization*, vol. 42 (Summer 1988), p. 434.

2. Warwick J. McKibbin and Peter J. Wilcoxen, "Climate Change after Kyoto: A Blueprint for a Realistic Approach," *Brookings Review*, vol. 20 (Spring 2002), p. 7.

3. Neil Carter, *The Politics of the Environment: Ideas, Activism, Policy* (Cambridge University Press, 2001), pp. 305–07.

4. Ute Collier and Ragnar E. Lofstedt, "Think Globally, Act Locally? Local Climate Change and Energy Policies in Sweden and the United Kingdom," *Global Environmental Change*, vol. 7, no. 1 (1997), p. 25. This article suggests a domestic policymaking process in Europe that may be analogous to that revealed in the United States, in which local or state units of government in Europe appear far more capable of taking early reduction initiatives than their national-level counterparts. In fact, this trend may only be accelerated as individual EU nations begin to move from more unitary governmental traditions toward a more American-style approach, whereby centralized and decentralized units share power.

5. Emma Daly, "Europeans Lagging in Greenhouse Gas Cuts," *New York*

Times, May 7, 2003, p. A10; "EU Slips in GHG Reduction Efforts," *Environmental Finance*, vol. 3 (May 2002), p. 12.

6. Carter, *Politics of the Environment*, p. 305.

7. The policy analyst Michael Grubb and his colleagues have written that "the United States has long argued that because of its culture and internal political structures, it would be held accountable to any specific commitments made—a situation it contrasted with what it perceives as the tendency of many European countries to declare fine targets without any detailed plan to achieve them." Michael Grubb, with Christiaan Vrolijk and Duncan Brack, *The Kyoto Protocol: A Guide and Assessment* (London: Royal Institute of International Affairs, 1999), p. 54.

8. Thomas C. Schelling, "The Cost of Combating Global Warming: Facing the Tradeoffs," *Foreign Affairs*, vol. 76 (November–December 1997), p. 13.

9. R. Kent Weaver, *Ending Welfare as We Knew It* (Brookings, 2001).

10. Eric Siy, Leo Koziol, and Darcy Rollins, *The State of the State Report* (San Francisco: Renewable Resource Institute, 2001), p. 23.

11. Russell Gold and Robert Gavin, "Fiscal Crises Force States to Endure Painful Choices," *Wall Street Journal*, October 7, 2002, p. A1; Andrew Caffrey and Russell Gold, "Governor, Get a Grip!" *Wall Street Journal*, November 1, 2002, p. B1.

12. Terry Davies, with Aracely Alicea, Robert Hersh, and Ruth Greenspan Bell, *Reforming Permitting* (Washington: Resources for the Future Press, 2001), p. 59.

13. For an excellent summary of this case and the larger dynamic of state regulation in a national marketplace, see John Dernbach and the Widener University Law School Seminar on Global Warming, "Moving the Climate Change Debate from Models to Proposed Legislation: Lessons from State Experience," *Environmental Law Reporter*, vol. 30 (November 2000), pp. 10977–99.

14. Jonathan Walters, "Save Us from the States!" *Governing*, June 2001, p. 20.

15. Quoted in "Industry Perspectives," in Environmental Council of the States, *Report of the ECOS Climate Change Workshop* (Washington, 1998), p. 8.

16. John Fialka, "States Protest Bush's Plan for Siting Power Lines," *Wall Street Journal*, May 15, 2001, p. A12. For an excellent analysis of transmission issues, as well as larger issues concerning electricity generation and distribution, see Timothy J. Brennan, Karen L. Palmer, and Salvador A. Martinez, *Alternating Currents: Electricity Markets and Public Policy* (Washington: Resources for the Future Press, 2002).

17. Tom Arrandale, "Tilting toward Windmills," *Governing*, June 2001, p. 46.

18. Rebecca Smith, "Overhaul of Energy Transmission Is Mired in Politics," *Wall Street Journal*, January 14, 2003, p. B2.

19. Michael S. Greve, *Real Federalism: Why It Matters, How It Could Happen* (Washington: American Enterprise Institute Press, 1999), p. 76.

20. The legislation specifically called upon the Minnesota PUC to "quantify and establish a range of environmental costs associated with each method of electricity generation. A utility shall use the values established by the commission in conjunction with other external factors, including socioeconomic costs, when

evaluating and selecting resource options in all proceedings before the commission, including resource plan and certificate of need proceedings." Minnesota, Statute 216B.2411 (October 12, 1993), chap. 356, sec. 3, pp. 8–9. Staff from the PUC and the state Pollution Control Agency subsequently developed a range of interim values for six emissions: sulfur dioxide, nitrogen oxides, lead, volatile organic compounds, particulates, and carbon dioxide. After an extended series of public hearings and a review by a Minnesota administrative law judge, the PUC voted in January 1997 to accept a range of $.30 to $3.10 a ton as the estimated environmental cost of carbon dioxide. This estimate was based on 1995 dollars, and values are to be updated using the gross national product price-deflator index as data become available. Minnesota, Rule E-999/CI-00-1636 (May 3, 2001).

21. Quoted in Jeffrey Ball, "U.S. Joins Fight against California Clean-Air Effort," *Wall Street Journal*, October 10, 2002, p. A2.

22. Martha A. Derthick, *Up in Smoke* (Washington: Congressional Quarterly Press, 2002); Alan Greenblatt, "The Avengers General," *Governing*, May 2003, pp. 52–56.

23. Letter from the Attorneys General of the State of New York, State of Connecticut, State of Maine, Commonwealth of Massachusetts, State of New Jersey, State of Rhode Island, and State of Washington to The Honorable Christine Todd Whitman, Administrator, Environmental Protection Agency, February 20, 2003, p. 1; available upon request from U.S. Environmental Protection Agency.

24. Jon Reisman, "N.E. Governors' Kyoto Accord with Canada Would Violate U.S. Constitution," *(Manchester, N.H.) Union Leader,* July 26, 2002, p. A19.

25. For two sobering assessments of Kyoto, see David G. Victor, *The Collapse of the Kyoto Protocol and the Struggle to Slow Global Warming* (Princeton University Press, 2001); Warwick J. McKibbin and Peter J. Wilcoxen, *Climate Change Policy after Kyoto: Blueprint for a Realistic Approach* (Brookings, 2002).

26. Charles O. Jones, *Clean Air* (University of Pittsburgh Press, 1975).

27. Frank R. Baumgartner and Bryan D. Jones, *Agendas and Instability in American Politics* (University of Chicago Press, 1993), p. 232. Similarly, Henry Butler and Jonathan Macey have noted that "there can be little doubt that federal policy would be better informed if it could draw on the divergent experiences of the states in dealing with other environmental problems." Henry N. Butler and Jonathan R. Macey, *Using Federalism to Improve Environmental Policy* (Washington: American Enterprise Institute Press, 1996), p. 46.

28. Sarah A. Binder, *Stalemate: Causes and Consequences of Legislative Gridlock* (Brookings, 2003).

29. Donald F. Kettl, "A Long Way from Austin," *Governing*, April 2003, p. 14.

30. On these various forms of diffusion and their interrelationships, see Karen Mossberger, *The Politics of Ideas and the Spread of Enterprise Zones* (Georgetown University Press, 2000).

31. Minnesota, *Environmental Regulatory Innovations Act*, Statute 114 C.01 (1996), sec. 1, p. 1.

32. Barry G. Rabe, "Permitting, Prevention, and Integration: Lessons from the States," in Donald F. Kettl, ed., *Environmental Governance* (Brookings, 2001), p. 53; Barry G. Rabe, "Toward the Sustainable State: Environmental Policy Innova-

tion in Minnesota," *Journal of Great Lakes Law, Science, and Policy,* vol. 1 (Fall 1998), pp. 191–210; Davies et al., *Reforming Permitting,* pp. 48–49.

33. Dernbach and the Widener University Law School Seminar on Global Warming, "Moving the Climate Change Debate," p. 10979.

34. Christopher H. Foreman Jr., "The Civic Sustainability of Reform," in Donald F. Kettl, ed., *Environmental Governance* (Brookings, 2001), pp. 159–60.

35. On the precedents for establishing such coalitions to support legislation, see Charles O. Jones, *The Presidency in a Separated System* (Brookings, 1995).

36. John Gummer and Robert Moreland, "European Union: A Record of Five National Protocols," in Pew Center on Global Climate Change, *Climate Change: Science, Strategies, and Solutions,* ed. Eileen Claussen (Leiden, Netherlands: Brill Academic, 2001), p. 92.

37. Grubb, with Vrolijk and Brack, *Kyoto Protocol,* chap. 3.

38. Ibid., p. 86.

39. Government of Canada, *Climate Change Plan for Canada* (Ottawa, 2002).

40. Pietro S. Nivola, "Energy Independence or Interdependence? Integrating the North American Energy Market," *Brookings Review,* vol. 20 (Spring 2002), p. 26.

41. Peter Hakim and Robert E. Litan, introduction to *The Future of North American Integration: Beyond NAFTA* (Brookings, 2002), pp. 15–16.

42. Nivola, "Energy Independence or Interdependence?" p. 27.

43. David B. Walker, *The Rebirth of Federalism* (New York: Chatham House, 2000); Timothy Conlan, *From New Federalism to Devolution: Twenty-Five Years of Intergovernmental Reform* (Brookings, 1998).

44. The National Academy of Public Administration has been a particularly strong champion of this concept through a series of reports on environmental governance published between 1995 and 2000. As a 1995 report concludes, "EPA and Congress need to hand more responsibility and decision-making authority over to the states and localities. A new partnership needs to be formed, one based on 'accountable devolution.'" A 1997 report further endorses this approach, noting that "EPA should reward good state performance with additional autonomy and flexibility" and highlighting the need to give states more freedom in the use of federal grants, the implementation of existing programs, and the development of new initiatives. See National Academy of Public Administration, *Resolving the Paradox of Environmental Protection* (Washington, 1997), pp. 203 and 208–09.

45. Quoted in Walters, "Save Us from the States!" p. 24.

46. According to the National Academy of Public Administration, "The principle is simple: one size does not fit all. Those states that are capable and willing to take over functions from the federal government should have full operational responsibility. In these cases, EPA should stay out of the way, with no second-guessing. In those states without the capability or the political will to assume responsibility, EPA should continue to exercise intensive oversight. And for those states falling between the two extremes, EPA should try to enhance their capabilities and help move them toward full delegation." National Academy of Public Administration, *Setting Priorities, Getting Results: A New Direction for EPA* (Washington, 1995), p. 2.

47. For a more detailed discussion of how a permit-based system might oper-

ate in the United States, see McKibbin and Wilcoxen, *Climate Change Policy after Kyoto*.

48. George Meyer, interview by author, Madison, Wisconsin, April 5, 2002.

49. Robert C. Shinn Jr., interview by author, Hainesport, New Jersey, April 11, 2002.

50. Michael Grunwald, "Everybody Talks about States' Rights. . . . But When It Comes to Acting on It in Congress, the Idea Is an Orphan," *Washington Post National Weekly Edition*, November 1, 1999, p. 29. One of Congress's strongest advocates of devolution, Ohio senator and former governor George V. Voinovich, concedes the difficulty of launching a serious discussion on this matter. "Everybody up here is constantly saying we should send power out of Washington, but we hardly ever do. I keep trying to get that across to people. It's just impossible to get anyone to listen." For a scholarly interpretation of the difficulties of matching devolutionary rhetoric with policy, see Walker, *Rebirth of Federalism*.

51. This tendency was clearly an impediment to active state engagement in the NEPPS program, particularly efforts to encourage states to apply for "leadership-track" status that offered maximum benefits. According to the policy analyst Shelley Metzenbaum, "Many states hesitated to be labeled in any way. Some states feared that a leadership label would provide ammunition to anti-environmental critics in state legislatures seeking to cut environmental agency budgets. At the same time, they feared that failure to earn the label would turn into a bad news story." Shelley Metzenbaum, *Making Measurement Matter: The Challenge of Building a Performance-Focused Environmental Protection System* (Brookings, 1998), p. 79.

Index

Abraham, Spencer, 14
Acid rain, 78, 99–100. *See also* Sulfur dioxide
Adams, Richard, 68
Agenda *21* Principles, 16
Agricultural issues: carbon credits, 32; carbon sequestration, 21, 67–73, 83, 151, 169–70; economic factors, 67–68
Agriculture, Department of, 68
Alabama, 20
Alaska, 8
Alberta, 174
Arctic National Wildlife Refuge, 146
Argentina, 4
Arizona, 20, 185n3
Arrandale, Tom, 78
Association of Local Air Pollution Control Officials, 80
Atlanta, 63–67
Audubon Society, 82
Australia, 52
Austria, 149
Automobile industry: air emissions, 17, 41, 62, 65, 141–42, 159; environmental protection and, 44; fuel efficiency, 15, 41, 62, 150; green-

house gases, 21; low-emission vehicles, 12, 150, 164. *See also* Michigan; Transportation industry

Baumgartner, Frank, 167
Bayer, Judith, 159
Belgium, 149
Berlin Mandates, 12–13
Blueprint for Intelligent Growth (N.J.), 140–41
Bluewater Network (Calif.), 143
Boyle, Robert, x, xi
Bradley, Jeb, 82–83
Brehm, John, 26
Brenner, John, 71, 72
Bush, George H. W., 10, 177
Bush (George H. W.) administration: climate change program, 9, 14; global climate change, 8; Energy Department, 81; UN Framework Convention on Climate Change, 3, 10
Bush, George W.: as governor of Texas, 1, 6, 50, 58; Kyoto Protocol and, 14; new federalism, 168; as presidential candidate, 2–3, 14, 34, 39, 183n16. *See also* Texas